ANIMAL AND PLANT
Anatomy

VOLUME CONSULTANTS

• Roger Avery, Bristol University, England • Amy-Jane Beer, Natural history writer and consultant
• John Friel, Cornell University, NY • Chris Mattison, Natural history writer and researcher
• Richard Mooi, California Academy of Sciences, San Francisco, CA • Kieran Pitts, Bristol University, England
• Adrian Seymour, Bristol University, England

9

Skeletal system – Trout

 **Marshall Cavendish**
Reference
New York

CONTRIBUTORS

Roger Avery; Richard Beatty; Amy-Jane Beer; Erica Bower;
Trevor Day; Erin Dolan; Bridget Giles; Natalie Goldstein;
Tim Harris; Christer Hogstrand; Rob Houston; John Jackson;
Tom Jackson; James Martin; Chris Mattison; Katie Parsons;
Ray Perrins; Kieran Pitts; Adrian Seymour; Steven Swaby;
John Woodward.

CONSULTANTS

Barbara Abraham, Hampton University, VA; Glen Alm,
University of Guelph, Ontario, Canada; Roger Avery, Bristol
University, England; Amy-Jane Beer, University of London,
England; Deborah Bodolus, East Stroudsburg University, PA;
Allan Bornstein, Southeast Missouri State University, MO;
Erica Bower, University of London, England; John Cline,
University of Guelph, Ontario, Canada; Trevor Day, University
of Bath, England; John Friel, Cornell University, NY; Valerius
Geist, University of Calgary, Alberta, Canada; John Gittleman,
University of Virginia, VA; Tom Jenner, Academia Británica
Cuscatleca, El Salvador; Bill Kleindl, University of Washington,
Seattle, WA; Thomas Kunz, Boston University, MA; Alan
Leonard, Florida Institute of Technology, FL; Sally-Anne
Mahoney, Bristol University, England; Chris Mattison; Andrew
Methven, Eastern Illinois University, IL; Graham Mitchell,
King's College, London, England; Richard Mooi, California
Academy of Sciences, San Francisco, CA; Ray Perrins, Bristol
University, England; Kieran Pitts, Bristol University, England;
Adrian Seymour, Bristol University, England; David Spooner,
University of Wisconsin, WI; John Stewart, Natural History
Museum, London, England; Erik Terdal, Northeastern State
University, Broken Arrow, OK; Phil Whitfield, King's College,
University of London, England.

Marshall Cavendish

99 White Plains Road
Tarrytown, NY 10591–9001

www.marshallcavendish.us

© 2007 Marshall Cavendish Corporation

Library of Congress Cataloging-in-Publication Data

Animal and plant anatomy.

   p. cm.
  ISBN-13: 978-0-7614-7662-7 (set: alk. paper)
  ISBN-10: 0-7614-7662-8 (set: alk. paper)
  ISBN-13: 978-0-7614-7673-3 (vol. 9)
  ISBN-10: 0-7614-7673-3 (vol. 9)
  1. Anatomy. 2. Plant anatomy. I. Marshall Cavendish
Corporation. II.
Title.

  QL805.A55 2006
  571.3--dc22

                         2005053193

Printed in China
09 08 07 06 1 2 3 4 5

MARSHALL CAVENDISH

Editor: Joyce Tavolacci
Editorial Director: Paul Bernabeo
Production Manager: Mike Esposito

THE BROWN REFERENCE GROUP PLC

Project Editor: Tim Harris
Deputy Editor: Paul Thompson
Subeditors: Jolyon Goddard, Amy-Jane Beer, Susan Watts
Designers: Bob Burroughs, Stefan Morris
Picture Researchers: Susy Forbes, Laila Torsun
Indexer: Kay Ollerenshaw
Illustrators: The Art Agency, Mick Loates, Michael Woods
Managing Editor: Bridget Giles

Contents

Skeletal system

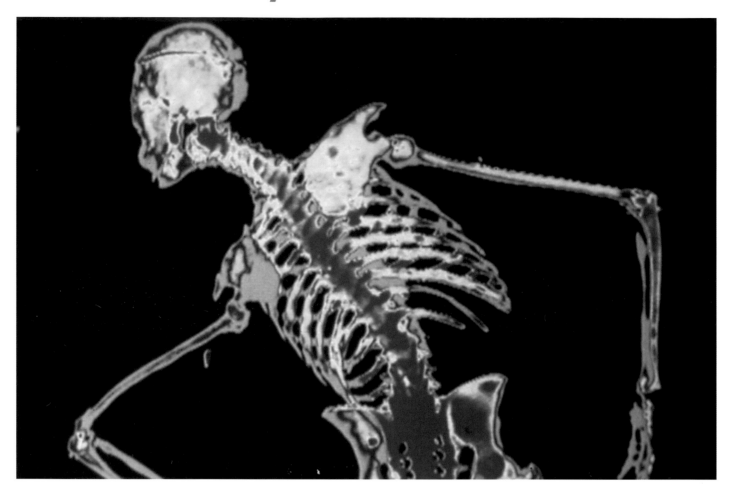

▲ *The tough, flexible human skeleton has about 206 bones of various shapes and sizes. This image shows the bones of the skull, vertebral column, ribs, pelvic girdle, and arms.*

Think of skeletons, and it is difficult to escape the concept of something dead. The supportive structures—bones and other hard parts, like shells—of many animals are the most enduring components of the body and often survive long after the animal is dead, when the flesh and other soft tissues have rotted away. However, in a living organism skeletal structures can be far from inert. They are very much living tissues, which change in response to use and within which a number of vital biochemical processes take place.

The skeleton is the part of an organism's body that gives it strength, rigidity, and support. In conjunction with muscle it makes precise movement possible, and in most animals the skeleton offers protection for other soft, vulnerable parts of the body such as the brain, heart, lungs, and other vital organs.

FEATURED SYSTEMS

Evolution of the skeleton

Life began in the sea. Early organisms were tiny: they were single cells surrounded by a membrane that held together all the vital components of life in gelatinous soup called the cytosol. In most such organisms, the membrane was reinforced with a network of microscopic fibers and an underlying scaffold of tubules. These networks gave cells a means of support and helped generate movement—the creeping of amoebas and the swimming of other protists. In this sense, the fibers and tubules were forerunners of both muscular and skeletal systems.

However, true skeletons did not evolve for a very long time. For much of the history of life on earth, organisms had no rigid support structures of any kind. Life-forms were fully aquatic and relied on water to support their bodies. Water is about 800 times more dense than air, so the organisms could function without a rigid skeleton. The fossil record for these early life-forms is very sparse precisely because they had very few hard parts that could be preserved.

This soft-bodied state of affairs lasted millions of years. In the absence of skeletons or any other hard parts, animal life nevertheless achieved an amazing level of diversity by 550 million years ago. However, even with water for support, a soft body has limitations, most notably a lack of power and speed. These are attributes that create a huge advantage for predators and prey alike, but they are

dependent on muscle power. And for muscles to operate efficiently and forcefully, they need something rigid to pull against.

About 500 million years ago, there was a period of great expansion in the number and variety of life-forms on Earth. During this period, called the Cambrian explosion, some animals developed hard skeletons that allowed them to maintain their body shape and move quickly. That offered a big advantage to predatory animals—and to prey animals trying to flee from predators. Thus skeletons and other support structures played an important part in a kind of evolutionary arms race. Once skeletons appeared, they evolved rapidly. Thus

▼ CROSS SECTION OF THE DIATOM NAVICULA
All the living matter of a diatom—a type of unicellular organism— is contained within the frustule, a rigid exoskeleton that is impregnated with silica.

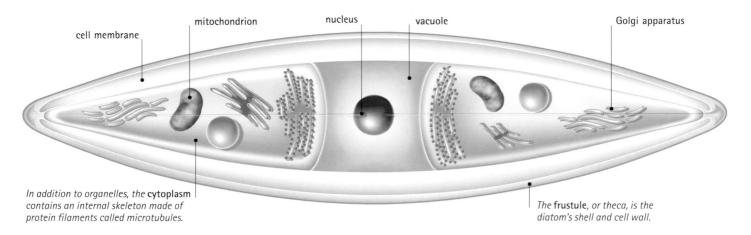

cell membrane | mitochondrion | nucleus | vacuole | Golgi apparatus

In addition to organelles, the **cytoplasm** *contains an internal skeleton made of protein filaments called microtubules.*

The **frustule,** *or theca, is the diatom's shell and cell wall.*

CONNECTIONS

COMPARE the skeleton of a **COELACANTH** with that of a **CROCODILE**. Similarities suggest how the skeleton of lobe-finned fish, such as the coelacanth, was modified for walking on land.

COMPARE the skeleton of an **OSTRICH** with that of an **EAGLE**. The ostrich evolved from a flying bird, so its forelimbs are more like those of birds than like those of other vertebrates that walk on two legs.

the transition from virtually no fossils in rocks more than 550 million years old to rocks full of an incredible diversity of fossils occurred relatively quickly.

The animals of the Cambrian explosion include representatives of most of the major animal groups that are extant (alive) today, plus many that have since become extinct. Alongside the ancestors of echinoderms, mollusks, and arthropods were small creatures with a long, bilaterally symmetrical body that were to give rise first to fish and then to other complex vertebrates.

The first hint of a skeleton in these simple pre-vertebrates was a long, sausage-shape structure called the notochord running along the animal's back. It was encased in a tough sheath that gave it rigidity and helped hold the body in shape, preventing the softer tissues from overstretching or being telescoped. The fish with the most primitive features alive today (hagfish and lampreys) still depend on a notochord for support, and the same structure is present in the embryos of all vertebrates. As

fish evolved, the notochord came to be replaced by a backbone made of articulated cartilage or bone structures. Having evolved the ability to produce these sturdy skeletal materials, fish evolved to provide support and protection elsewhere in the body: in the head and fins.

From fins to limbs

Vertebrates with limbs instead of fins are called tetrapods. Tetrapod means "four feet," though the description applies to many animals in which the feet are modified into something else, such as flippers or wings; and to some that have no legs or feet at all, such as snakes.

Scientists agree that the limbs of tetrapods evolved from the fins of certain types of fish called sarcopterygians. There is less agreement about exactly how this development took

▼ Priscacara liops *is an extinct lobe-finned fish that lived in North America 50 million years ago. All land vertebrates evolved from an ancestral lobe-finned fish, with the bones of the fish's fleshy pelvic and pectoral fins developing into the limb bones.*

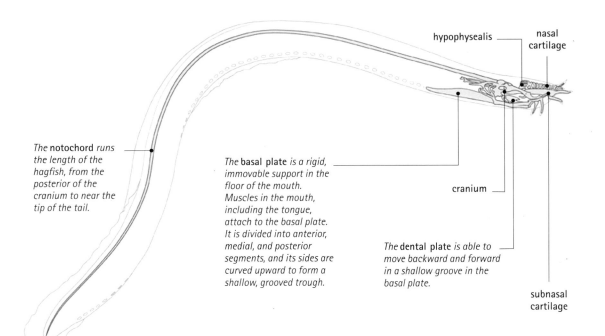

hypophysealis

nasal cartilage

cranium

The notochord runs the length of the hagfish, from the posterior of the cranium to near the tip of the tail.

The basal plate is a rigid, immovable support in the floor of the mouth. Muscles in the mouth, including the tongue, attach to the basal plate. It is divided into anterior, medial, and posterior segments, and its sides are curved upward to form a shallow, grooved trough.

The dental plate is able to move backward and forward in a shallow groove in the basal plate.

subnasal cartilage

◀ **Atlantic hagfish**
The skeleton of a hagfish is made of cartilage, not bone. Most of the elements of the skeleton are in the head. Unlike that of other adult vertebrates, the spinal cord of a hagfish is protected not by bony vertebrae but by a flexible notochord. A hagfish lacks ribs.

place, but the sequence of events was probably something like this. An ancestral fish had two pairs of lobed fins, like those of modern-day lungfishes or coelacanths. The ancestor was able to use these fins as levers to help pull itself along the sea floor. In shallow water, where the fish's body weight was not fully supported by the water, the technique was more effective than conventional swimming, and by pressing its fins into the ground as it wriggled from side to side the fish was even able to "swim" on land. At first, the fin-limbs did not carry much weight, but as fish found new advantages to excursions onto land (finding food and avoiding predators), natural selection began to favor those with longer and stronger limbs and pectoral and pelvic girdles (shoulders and hips) sturdy enough to take the strain. Thus evolved the first amphibians.

Taking the strain

The backbone of a fish acts to maintain the shape of the body, in particular resisting compressive forces and keeping the animal from telescoping or overstretching. In most species of fish, the backbone also has long spines for the attachment of the large trunk muscles that power swimming. However, since terrestrial vertebrates move on four legs without the support of water, their backbone also has to take on a weight-bearing role.

Different branches of the vertebrate family tree have adopted different approaches to the problem of moving and supporting the body on land. Salamanders have stuck with the original solution. Their skeleton has many similarities to that of their fish ancestors, and they retain a fishlike side-to-side wriggle when swimming and when moving on land. A salamander's legs are short and weak. When the salamander is resting, the weight of the body is lowered onto the ground. Lifting the body creates enormous stress on the legs and the long backbone, and thus walking is tiring and inefficient for these animals. In frogs and toads, walking is much less difficult, mainly because the body is much shorter. A typical frog has just nine relatively large vertebrae, and they interlock closely to form a rigid support with virtually no side-to-side flexibility. The sacral vertebrae are fused to the pelvic girdle, which is strengthened by a rigid central rod of bone,

EVOLUTION

The first tetrapod?

The earliest known tetrapod is an amphibian called *Ichthyostega*, which lived about 360 million years ago. Apart from the limbs, the skeleton of *Ichthyostega* is very much like that of a fish. *Ichthyostega* would have spent much of its time in water, but it was also able to move (and breathe) on land.

Archaeopteryx: A missing link

There is no doubt that birds evolved from reptiles. However, when fossil hunters discovered the proof, it surpassed all expectations. *Archaeopteryx* is a classic example of a missing link—an extinct organism showing a blend of characteristics from two groups and demonstrating how one group might have evolved from the other. *Archaeopteryx* had the skeleton of a reptile, with a long tail, small sternum (breastbone), small coracoid, and five-fingered forelimbs. But clearly, it also had feathers.

▼ *Many scientists believe that birds evolved from dinosaurs about 150 million years ago. This is the most complete fossil of* Archaeopteryx, *which was found in Solnhofen limestone deposits in Germany in 1880 and is now housed in the Humboldt Museum, Berlin, Germany.*

Toe count

Fossil evidence shows that early tetrapods had a variable number of digits on each limb: for example, eight in *Acanthostega*, seven in *Ichthyostega*, and six in *Tulerpeton*. However, in the evolutionary line that gave rise to more advanced vertebrates, five digits were the norm. The hands, feet, flippers, wings, and paws of all living amphibians, reptiles, birds, and mammals are derived from the same five-fingered structure, known as the pentadactyl limb.

Reptiles

Reptiles show the greatest skeletal diversity of any group of vertebrates. They have varied in size from tiny 1-inch-long geckos to colossal dinosaurs such as *Diplodocus* and *Brachiosaurus*, which stood up to 60 feet (18 m) tall and weighed 60 tons (61 metric tonnes). To put this into perspective, you could crush the whole skeleton of the smallest lizard with one toe, but just one neck bone from the largest known dinosaur could do the same to you!

Reptile skeletons vary in form as well as size. Faced with the skeletons of a large turtle and a python, it is difficult to see how one, with its short body, beaklike jaw, flippers, and shieldlike carapace, can possibly be related to the other, with its loose, spindly jaw, enormously long backbone, hundreds of ribs, and absence of functional limbs.

Mammals and birds

Compared with the very great diversity of the reptiles, the skeletons of mammals and birds, both of which groups are descended from reptilian ancestors, are relatively uniform. The body plan of birds, in particular, is restricted by the requirements of flight. Even flightless birds like the ostrich and kiwi are constrained by this, since they evolved from ancestors that flew. Birds have fewer skeletal elements than either reptiles or mammals because many bones are fused to provide the strength and rigidity needed for flight.

the urostyle, derived from fused caudal vertebrae. The pelvic girdle of frogs is very large because it connects the backbone to the extremely long hind legs.

Plant support structures

We tend not to think of plants as having skeletons, but they do have support structures. Often these structures are so effective that they last for hundreds of years. The skeletal material produced by the largest vascular plants—the wood of trees—is equally useful to humankind for construction and as fuel.

Nonanimal life, just like that of animals, began in the seas with the appearance of single-cell organisms about 1 billion years ago. Over time, some of these very simple life-forms gave rise to colonial and multicellular species, which had the potential for division of labor and specialization of different cell types. As with animals, plants had little need of rigid support as long as they remained in the water, and even after they encroached onto land many stuck with a simple form that simply encrusted or draped over existing structures and required little support. For aquatic plants, however, there are big advantages to growing upward, reaching toward the sun and pushing reproductive structures into the air where spores and seeds are more likely to be dispersed by the wind.

Plant skeletons evolved in tandem with the internal transport systems of vascular plants. In fact, to a large extent skeleton and vascular systems are one and the same thing. The branching skeleton seen in the leaves of vascular plants is a network of tiny xylem fibers and phloem tracheids that carry water and nutrients to every part of the plant. The vessels extend along every stem and twig, through the trunk of trees and into the roots. Because the walls of these tubes are reinforced to keep them open, they offer support to the plant as a whole. The supporting role of tracheids is enhanced by a thick layer of cells and associated material called the cambium. The cambium of woody plants is particularly large, and contains fibers of complex carbohydrates called cellulose and hemicellulose, held together with a special protein called lignin.

CLOSE-UP

Tendrils

Many climbing plants, such as sweet peas, cucumbers, and passionflowers, have threadlike structures called tendrils for support and attachment. Depending on the species of plant, tendrils can be modified leaves, stems, or petioles (leaf stalks). On contact with a solid object, such as a branch, a tendril twirls around it—a phenomenon known as thigmotropism. The cells that make contact with the object lose water and become smaller than the outer cells. In this way, the tendril curves around the object, supporting the plant as it grows upward.

▼ TRUNK CROSS SECTION
Apple tree
The outside layer of the trunk is nonliving bark, which provides a support structure for the tree. Beneath the "skeleton" of the trunk is the cambium, which transports water and lays down new xylem and phloem tissue.

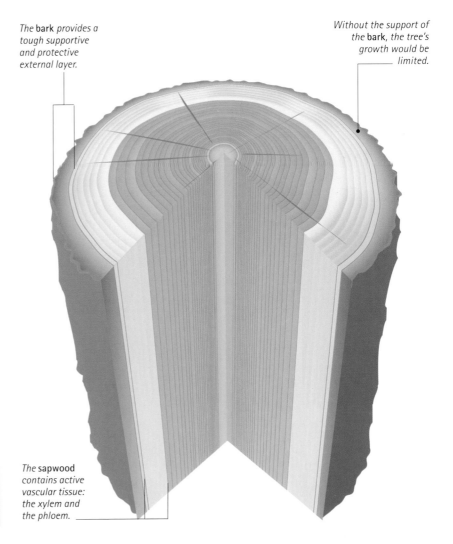

The bark provides a tough supportive and protective external layer.

Without the support of the bark, the tree's growth would be limited.

The sapwood contains active vascular tissue: the xylem and the phloem.

Invertebrate support structures

CONNECTIONS

COMPARE the exoskeleton of the **WEEVIL** with that of the **CRAB**. The weevil cuticle is less mineralized than that of the crab because it does not have access to such a ready supply of calcium carbonate.

COMPARE the hydraulic support system of the **EARTHWORM** with that of the **SEA ANEMONE**. These animals are not related, but both use water pressure to give their body rigidity and to move about.

Even the very simplest animals can have skeletons. Sponges are the simplest form of multicellular animal life. The body has just two layers (compared with three in most multicellular animals). Between the inner and outer surfaces of the sponge body is a mass of supporting material called mesohyl. In most sponges, this is a gel-like mass containing various stiffening elements such as protein fibers (collagen and spongin) and mineral spicules. The stiffening elements are sometimes preserved when the animal dies, leaving a delicate, brittle skeleton that looks like fine lace.

Even animals that have no hard body parts must be able to hold their body shape. Many soft-bodied organisms use water pressure to achieve this. To understand how water can help give a structure form and rigidity, imagine an empty balloon. It contains no hard parts and on its own is flimsy and floppy. However, if the balloon is filled with water and sealed, it takes on a three-dimensional form. The more water it contains, the tauter the latex of the balloon is stretched and the more rigid the shape becomes. All living organisms gain some measure of support from the water they contain. Plants are no exception: they wilt and loose rigidity when dehydrated.

Water cannot be compressed. It resists being squeezed by flowing from the areas of greatest pressure to areas of less pressure. Just as a hard skeleton provides muscles with something to pull on, tissues that are firmly plumped up with water can offer the resistance required to generate muscular locomotion. Zoologists call this rigidity a hydraulic skeleton, and soft-bodied animals use it to move. By contracting

▶ The seawater-filled cavity (coelenteron) of jellyfish, such as the many-ribbed hydromedusa (right), and its gel-like body tissues form a supportive structure called a hydrostatic skeleton.

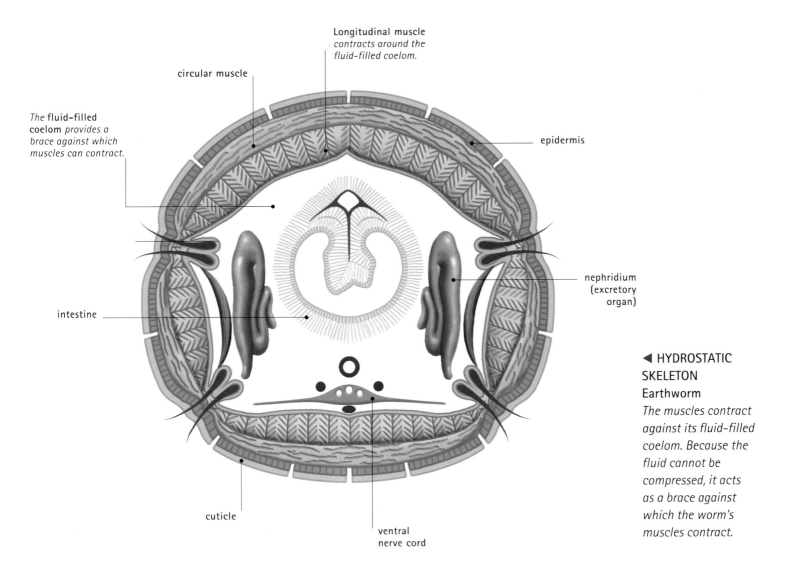

circular muscle

Longitudinal muscle
*contracts around the
fluid-filled coelom.*

The fluid-filled
coelom *provides a
brace against which
muscles can contract.*

epidermis

intestine

nephridium
(excretory
organ)

cuticle

ventral
nerve cord

◀ HYDROSTATIC
SKELETON
Earthworm
*The muscles contract
against its fluid-filled
coelom. Because the
fluid cannot be
compressed, it acts
as a brace against
which the worm's
muscles contract.*

muscles in one part of the body, an animal such as a sea anemone is able to force water into another area, such as a tentacle, which must then expand to accommodate the water.

The hydraulic skeleton is developed to best effect in the annelids, segmented worms such as earthworms, ragworms, and leeches. Each segment in the annelid body is effectively a bag of fluid wrapped in several thin layers of muscle. This arrangement allows the animal to move in a variety of ways. For example, when the muscle on one side of a segment contracts, the other side expands to accommodate the fluid. When this happens simultaneously in several segments, the whole body curves toward the side that is contracting. By alternating contractions the worm performs a characteristic wiggle that can propel it forward; that is how ragworms crawl and swim.

IN FOCUS

A change of armor

The rigid exoskeleton of crustaceans such as lobsters and crabs is not elastic, and in order to grow the animal must shed its protective covering and produce a new, larger one. The process may have to be repeated dozens of times during the animal's lifetime, and the process becomes increasingly arduous as the animal grows. Before shedding its armor, the animal becomes dehydrated, so that it shrinks a little inside its armor. It also reabsorbs minerals from the exoskeleton into the body. The exoskeleton becomes brittle and splits open, allowing the animal to crawl out. To begin with, the new exoskeleton is soft and crumpled, but as the animal swells back to normal size, it straightens out and begins to harden. The process can take several days, during which time the soft-bodied animal is very vulnerable to predation. The animal often eats its old exoskeleton to help replace lost protein and minerals.

Alternatively, if circular muscles cause a segment to constrict, hydraulic pressure forces it to lengthen. Thus a worm can make each segment (and its whole body) short and fat, or long and thin. By alternating from one state to the other, annelids like earthworms effect a gradual creeping type of locomotion.

The echinoderm test

Echinoderms are a group of marine animals including starfish, sea cucumbers, and sea urchins. Most sea cucumbers are soft-bodied, but all other echinoderms possess some kind of internal skeleton.

In starfish the skeleton is made up of a lattice of mineral rods, crosses, or plates embedded in a layer of connective tissue within the body wall. In species such as the common red sea star, the skeleton is relatively loose, whereas in other species it can be very highly mineralized and brittle. In sea urchins, the skeleton is even more developed; it is a regular jigsaw of plates

that form a roughly spherical or discus-shaped shell called a test. The plates of the test are covered with small bumps, to which the mobile spines are attached in the living animal. The plates are perforated by tiny holes through which the suckerlike tube feet can be extended when the animal crawls. The tube feet operate by water pressure, a little like the hydraulic skeletons of corals, sea anemones, and annelids.

The shells of mollusks

Mollusks are essentially soft-bodied animals that have developed a characteristic form of body armor. The shells, or valves, of mollusks such as mussels, clams, snails, and limpets are not really skeletons, but they protect the animal from physical attack and—in many tidal and land-dwelling species—from dehydration. The shells may also serve as a scaffold from which some species can suspend organs such as gills and filters, which lack their own rigidity and would not work if allowed to

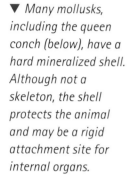

▼ *Many mollusks, including the queen conch (below), have a hard mineralized shell. Although not a skeleton, the shell protects the animal and may be a rigid attachment site for internal organs.*

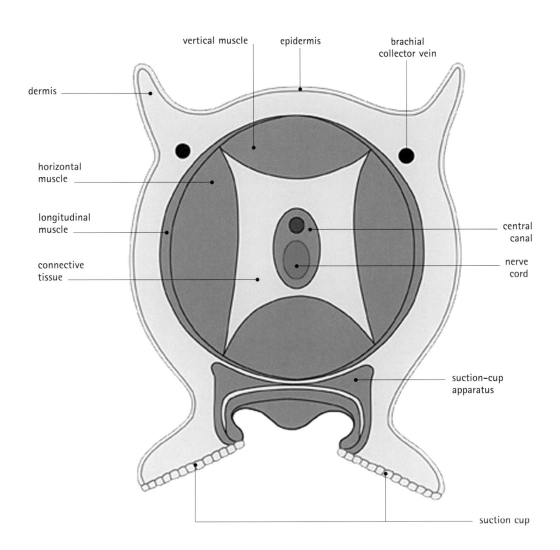

vertical muscle   epidermis

brachial collector vein

dermis

horizontal muscle

longitudinal muscle

connective tissue

central canal

nerve cord

suction-cup apparatus

suction cup

◄ CROSS SECTION OF OCTOPUS TENTACLE

An octopus does not have an internal skeleton. Its arm and sucker are formed from a complex arrangement of muscles, working against one another and enabling the octopus to wrap around and grip very tightly onto prey.

lie flat. In cuttlefish and squid the shell is internal; in octopuses it has disappeared.

The tissues of a mollusk's mantle—the soft part of the animal that contacts the shell directly—lay down the shells throughout the animal's lifetime. The shells of mollusks are made of alternating layers of protein and two different crystalline forms of calcium carbonate: calcite and aragonite. Marine mollusks have access to much more calcium carbonate than freshwater or land-dwelling species and thus can create heavier shells. The shiny material lining the shells is called nacre, or mother-of-pearl. It contains about 95 percent aragonite, but the bulk of the shell matrix in marine mollusks is usually calcite, which is more stable chemically in seawater.

CELL BIOLOGY

Arthropod cuticle

The exoskeleton of arthropods such as insects and spiders is made of protein and chitin, a polysaccharide secreted by the cells in the underlying skin. The cuticle is made of several layers: a thin, waxy, outer epicuticle; a pigmented exocuticle; and a thicker endocuticle. The exocuticle is missing from the jointed areas, making the exoskeleton flexible. In large crustaceans such as crabs and lobsters, the exocuticle and endocuticle also contain large quantities of minerals, especially calcium carbonate and calcium phosphate.

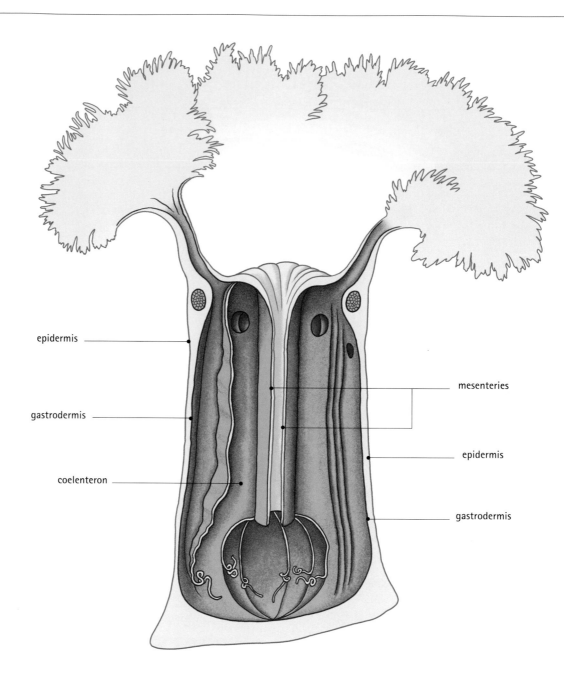

▶ **Common plumose sea anemone**
Two walls of cells—the epidermis and gastrodermis—provide a support structure around the animal's body cavity, or coelenteron. Internal walls called mesenteries give additional support.

epidermis

gastrodermis

coelenteron

mesenteries

epidermis

gastrodermis

Arthropod exoskeleton

The arthropods are the largest and most successful group of invertebrates and include insects, arachnids, centipedes, millipedes, and crustaceans. The group owes much of its remarkable success to its unique suit of armor, the exoskeleton. The exoskeleton consists of a cuticle made of the polysaccharide chitin and serves as both structural support and body protection. In marine arthropods, the cuticle is often reinforced with deposits of calcium carbonate and calcium phosphate. To allow movement, the cuticle is jointed at flexion points all round the body where the cuticle is thin and bendy. In "primitive" arthropods, such as wood lice, millipedes, and the extinct trilobites, the body plates follow the same pattern as the body segments, with each segment encased in four plates: the tergum on the back, the sternum on the underside, and two side plates called pleura.

In other groups of arthropods, such as insects and arachnids, the segmental arrangement is sometimes lost over part or all of the body, and the plates are either fused or subdivided. The legs and other appendages are sheathed in chitin tubes joined by articulating membranes that allow them to bend. The word *arthropoda* means "jointed feet."

Growing pains

One of the problems with living inside a rigid structure is that it restricts an organism's ability to grow. Arthropods get around this problem by shedding their exoskeleton periodically in a process called ecdysis, or molting. Mollusks and some single-cell organisms called foraminiferans molt by laying down more material at the edges of their shell, resulting in the ever-enlarging spiral of a snail shell or the concentric growth rings on the valves of mussels and scallops. For an echinoderm such as a sea urchin, in which the test must grow with the animal, shedding the test and adding material at the opening are not practical solutions. Instead, each of the plates in the test grows around its edges. Plates that are actively growing are only weakly attached to

Coral reefs

Coral polyps would be inconspicuous animals were it not for the enormous support structures they build. These amazing structures, which accumulate to form vast reefs, are more like houses than skeletons, but the stony material from which they are made is secreted by the tiny soft-bodied animals living within.

their neighbors. Thus the test of a sea urchin that dies during a period of growth falls apart easily. A sea urchin that has not grown much recently, perhaps because of a lack of food, has a stronger test in which the joints, or sutures, between the plates are heavily mineralized and firmly attached to one another.

▼ *A lobster's soft parts are enclosed in a tough cuticle, or exoskeleton, which provides support for the animal, points of attachment for muscles, and protection against predators. The cuticle of lobsters is made of chitin, a complex carbohydrate similar in structure to cellulose, bound up with various proteins. The cuticle is much thinner at the joints, to allow movement.*

Vertebrate skeleton

Bone is the predominant skeletal material in most adult vertebrates. Like cartilage, it develops from connective tissue and consists of isolated cells that secrete a resilient matrix material. Whereas the matrix of cartilage is usually springy and translucent, that of bone accumulates mineral salts, mainly calcium carbonate and calcium phosphate, which combine to form a material called hydroxyapatite. The mineralization of bone

CELL BIOLOGY

Osteoblasts and osteoclasts

The bones of vertebrates are laid down and sculptured throughout life by the actions of two types of bone cells. Osteoblasts secrete bone matrix and remain active even when an animal is fully grown, helping broken bones heal and replacing old bone material that is continually being destroyed by the other important type of bone cell, the osteoclasts. Osteoclasts digest bone matrix, allowing the constituent minerals to be recycled. In adult humans, about 10 percent of bone is replaced in this way each year, ensuring that the skeleton stays strong. In older women, there is a tendency for osteoclasts to demineralize bone faster than osteoblasts can replace it. This demineralization leads to a weakening of the bones—a disorder called osteoporosis.

▼ *This micrograph shows human osteoblasts. These cells are found in bone and secrete substances such as collagen that form bone matrix.*

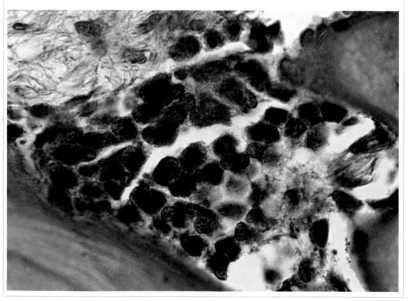

CLOSE-UP

Bone density

Compared with the bones of mammals, those of birds are about one-third the weight. Bird bones are hollow, with an internal scaffold of fine struts, making them strong but light. This characteristic is a big advantage for flight. The bones of some animals are unusually dense and heavy even when compared with mammals. Ocean-dwelling manatees, for example, have heavy bones that help counteract the buoyancy of their thick layer of fat. Heavy bones allow them to feed on underwater plants without having to expend energy to stay submerged. The density of bone is also related to its strength: in humans, the lower jaw is the hardest, densest bone, whereas in male deer the antlers are hardest and densest.

causes it to harden and become opaque. Since bone is impermeable to nutrients, it requires its own blood supply. Thus it is penetrated by a fine network of microscopic channels called Haversian canals that allow blood and lymph to pass to and from the cells embedded in the tissue. The center of a large bone is particularly richly supplied with vessels. There, the bone structure is relatively open and spongy-looking, with many small holes and channels. This spongy area also contains bone marrow, a fatty substance important in the production of blood cells.

The support structures of vertebrates are true skeletons; they are internal scaffolds made of two different kinds of tissues—bone and cartilage. Both types of tissues are derived from connective tissue, but they are very different materials.

Cartilage is a tough, flexible tissue that begins to form early in the development of an embryo. Cartilage performs a supporting role long before bones develop. Usually cartilage is a deep tissue: that is, it does not form close

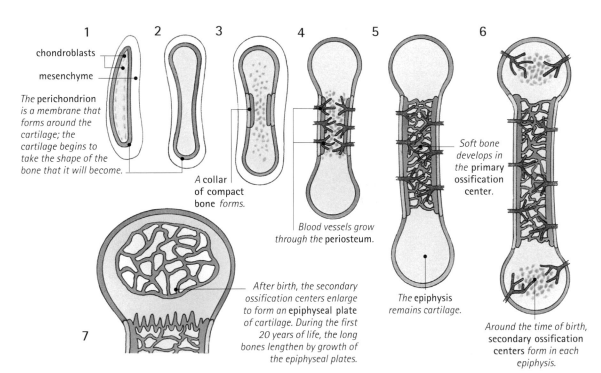

1

chondroblasts

mesenchyme

The perichondrion is a membrane that forms around the cartilage; the cartilage begins to take the shape of the bone that it will become.

2

3

A collar of compact bone forms.

4

Blood vessels grow through the periosteum.

5

The epiphysis remains cartilage.

Soft bone develops in the primary ossification center.

6

Around the time of birth, secondary ossification centers form in each epiphysis.

7

After birth, the secondary ossification centers enlarge to form an epiphyseal plate of cartilage. During the first 20 years of life, the long bones lengthen by growth of the epiphyseal plates.

◀ DEVELOPMENT OF LONG BONES IN A FETUS AND CHILD
Human
Fetal bones develop from tissue called mesenchyme. The chondroblasts are cartilage-making cells. Diagrams 1–5 show different stages as the embryo grows. 6 illustrates a long bone as it is around the time of birth, and 7 shows the bone of a young child.

◀ In addition to an internal skeleton, armadillos, such as this nine-banded armadillo, are covered by a "shell" or exoskeleton of bony horn-covered plates that protects the animal from predators.

under the skin or at the body surface. Exceptions to this general rule include the human nose and the external structure of the ears in most mammals.

Hyaline cartilage, a glassy type of cartilage found in joints, is a slightly elastic tissue in which the predominant compound is a polysaccharide called chrondromucoprotein. This translucent, smooth-textured material is a sort of set gel secreted by specialized cells called chondrocytes. The chondromucoprotein matrix is strengthened by many fibers crisscrossing through it. The chrondrocytes become trapped by the material that they secrete, ending up encapsulated in lacunae, small bubbles within the cartilage. These cells receive all the vital nutrients and oxygen they need to function by simple diffusion through the cartilage gel, since only large cartilaginous structures have a blood supply. Not all cartilage has the same springy texture. In vertebrates such as sharks, whose entire skeleton is composed primarily of cartilage, the tissue is so heavily calcified that it resembles bone.

Vertebrate skull

CONNECTIONS

COMPARE the skull of a **CHIMPANZEE** with that of a **CROCODILE**. The chimpanzee's skull has orbits in the front, so the eyes face forward and provide good binocular vision. The crocodile has a long snout, with the orbits high on the sides of the skull, giving good all-around vision. The chimpanzee's cranium, which houses a large brain, takes up a large proportion of the skull, whereas that of the crocodile is relatively small. The lower jaw of the crocodile is almost as large as the upper jaw, whereas the chimp's lower jaw is far smaller.

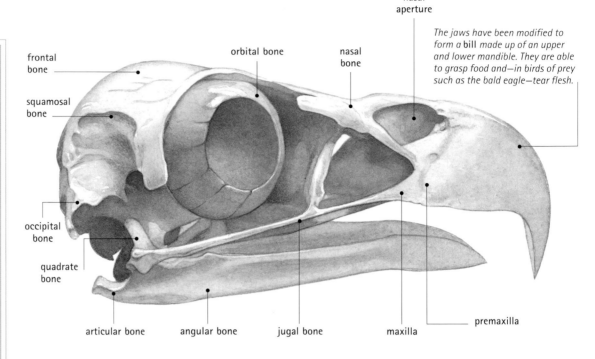

nasal aperture

frontal bone

orbital bone

nasal bone

*The jaws have been modified to form a **bill** made up of an upper and lower mandible. They are able to grasp food and—in birds of prey such as the bald eagle—tear flesh.*

squamosal bone

occipital bone

quadrate bone

articular bone   angular bone   jugal bone   maxilla

premaxilla

▲ **Bald eagle**
The skull of a bird is composed of thin sheets of bone with tiny supporting struts. The skull is extremely strong to support the jaw muscles and protect the brain. The jaws of a bird are modified to form a grasping bill. Eagles have a hooked upper mandible that is used to tear flesh.

In all vertebrates the brain is encased in a bony or cartilaginous box called the cranium, which is the most important element of the skull. Despite its appearance, the skull is not a single large bone but a structure formed from many smaller bones, which become fused during the development of the animal. This is more obvious in some animals than in others. In typical mammals, for example, the skull is made up of 34 bones. In a mature skull, the joints between the bones appear as fine, squiggly lines. In most vertebrates the skull can be divided into two main parts: the braincase, or cranium, which includes the facial bones;

and the part that in fish forms the gill arches and in other vertebrates forms the lower jaw.

The skull of both birds and reptiles has a characteristic articulation (joint) of the lower jaw made up of articular and quadrate bones. In mammals, the lower jaw consists of just one bone, the dentary. In the reptile ancestors of mammals, there were three pairs of accessory jawbones: the angular, articular, and prearticular bones. In mammals, these three pairs of bones have become the malleus, incus, and stapes of the inner ear. They are greatly reduced in size but perform an important role in transmitting the vibrations mammals interpret as sound. The acute hearing of birds is due in part to a different arrangement: vibrations are transmitted through a single auditory ossicle (ear bone) called the columella.

In all vertebrate embryos, the development of the skull follows a more or less similar pattern,

CLOSE-UP

Beaks

When the ancestors of birds evolved forelimbs devoted to powered flight, they had to sacrifice many of the other useful functions of forelimbs, such as manipulating food, grooming, signaling, fighting, crawling, climbing, and grasping. In compensation, birds use their jaws for a much greater range of functions than other tetrapods. The jaws of birds are encased in a sheath of horny material, the beak. Bird beaks have evolved an astounding array of shapes and sizes, allowing birds to perform a variety of highly specialized tasks.

Horns and antlers

The impressive headgear of hoofed animals such as cattle, rhinoceroses, and deer is formed largely from bone. Horns are permanent structures that grow from the frontal bones of all male and some female cattle, antelope, and related animals. Horns are sheathed in a layer of tough keratin, the same material that is found in hair and nails. Antlers are temporary structures grown and cast off annually by male deer; antlers have no horny covering.

several other newly formed cartilaginous structures until the brain is enclosed in a loose, symmetrical jigsaw of plates. At about the same time, a series of paired cartilaginous struts form in the region of the throat, beneath the skull. In fish, the struts become the gill support arches. The first of these arches, called the mandibular arch, progresses to form part of the upper and lower jaws, but these primary jaws persist only in sharks and rays. In most bony fish and higher vertebrates they are soon replaced by new bone derived from the overlying skin. The second set of cartilaginous structures form the hyoid arch, which in fish helps suspend the jaw from the main part of the skull. In more advanced vertebrates, the hyoid arch supports the base of the tongue and is used in vocalization.

In all vertebrates except cartilaginous fish, the cartilaginous structures of the embryonic skull are gradually replaced by bone. In large-brained mammals, and especially in humans, this process is not completed until after birth. Young babies have a soft spot, the fontanelle, on the top of the skull that allows the head to be compressed so it can pass through the birth canal.

which repeats the sequence of evolutionary development. The braincase develops from neural crest tissue at the back of the brain. First to appear are a pair of small cartilaginous structures, one on each side of the front end of the notochord. These grow and fuse around the notochord, forming a basal plate, which grows out around the brain, meeting and fusing with

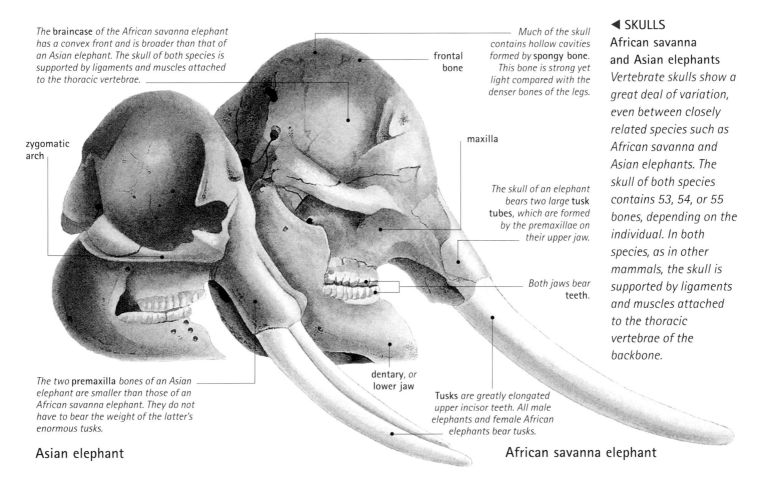

The **braincase** of the African savanna elephant has a convex front and is broader than that of an Asian elephant. The skull of both species is supported by ligaments and muscles attached to the thoracic vertebrae.

Much of the skull contains hollow cavities formed by **spongy bone**. This bone is strong yet light compared with the denser bones of the legs.

frontal bone

maxilla

The skull of an elephant bears two large **tusk tubes,** which are formed by the premaxillae on their upper jaw.

zygomatic arch

Both jaws bear teeth.

The two **premaxilla** bones of an Asian elephant are smaller than those of an African savanna elephant. They do not have to bear the weight of the latter's enormous tusks.

dentary, or lower jaw

Tusks are greatly elongated upper incisor teeth. All male elephants and female African elephants bear tusks.

Asian elephant

African savanna elephant

◀ SKULLS
African savanna and Asian elephants
Vertebrate skulls show a great deal of variation, even between closely related species such as African savanna and Asian elephants. The skull of both species contains 53, 54, or 55 bones, depending on the individual. In both species, as in other mammals, the skull is supported by ligaments and muscles attached to the thoracic vertebrae of the backbone.

Vertebrate axial skeleton

COMPARE the neck bones of the **GIRAFFE** with those of a **WILDEBEEST**. Both have seven cervical vertebrae, despite the difference in neck length.

COMPARE the coccyx of the **HUMAN** with the pygostyle of the **PENGUIN** and the urostyle of the **BULLFROG**. All are evolved from the longer tails of their ancestors.

Vertebrates are by definition animals in which the notochord becomes gradually replaced by bony or cartilaginous units called vertebrae, which collectively form the backbone, or spine. However, in the simplest of vertebrates, this substitution is incomplete, and the notochord remains the main support structure throughout life. Indeed, in hagfish, the only evidence of any backbone is a row of tiny cartilaginous structures in the tail, forming the most rudimentary of vertebrae. In the lamprey, there are cartilaginous vertebrae along the full length of the notochord, but they are still relatively small.

The vertebrae of sharks and rays are much more substantial. Each one develops around the notochord, which is constricted into a shape resembling a string of beads. The part of the vertebra through which the notochord passes is called the centrum. In bony fish and tetrapods, the centrum is filled in with bone during embryonic development, so the notochord is constricted more and more and eventually broken into a series of cushions between the vertebrae. These cushions are the intervertebral disks.

Vertebrae

In addition to their supportive role, the vertebrae also protect the spinal cord. This vital tract of nerve runs along the back of the animal, through a tube created by a series of arches, one on the top of each vertebra. In some vertebrates there is a second set of arches—the hemal arches—below the centrum. The vertebrae may also bear a number of spine projections, or processes, which serve as attachment points for various muscles. The processes vary considerably in size and shape at different points along the backbone and between species, so a zoologist can determine the difference between, for example, the cervical (neck) vertebrae and the lumbar (back) vertebrae of an animal.

▶ THORACIC VERTEBRA

Whale

Whales have 14 thoracic vertebrae in the trunk region. These vertebrae do not permit much movement, but the ribs attached to them (one on each side) are mobile. Each vertebra is large, with broad surfaces and extensions called neural spines and transverse processes. The processes act as attachment points for muscles, and the neural spines act as levers, increasing the power that can be transmitted by the muscles to make the spine flex.

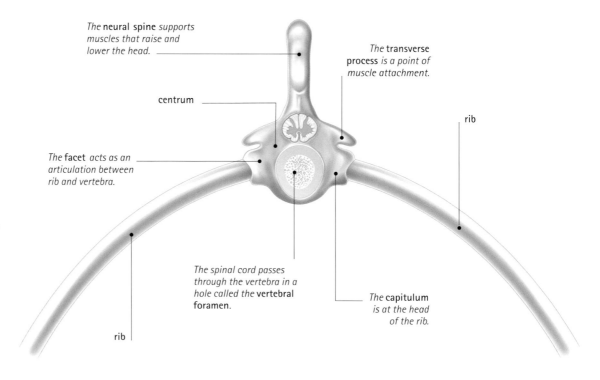

The **neural spine** *supports muscles that raise and lower the head.*

The **transverse process** *is a point of muscle attachment.*

centrum

rib

The **facet** *acts as an articulation between rib and vertebra.*

The spinal cord passes through the vertebra in a hole called the **vertebral foramen**.

The **capitulum** *is at the head of the rib.*

rib

The vertebrae at the top of the spine are called cervical vertebrae. The first of these, which articulates directly with the base of the skull, is the atlas. In mammals, the atlas has two depressions in the anterior (front) face nearest the skull; these fit snugly around two bumps in the base of the skull called occipital condyles. The bumps are two of the many skeletal features that set mammals apart from other vertebrates. In amphibians, the atlas is the only cervical vertebra. In "higher" vertebrates, the atlas articulates with the second cervical vertebra, or axis. Following the atlas and axis are a series of conventional cervical vertebrae, making a total of seven in most mammals but as many as 25 in some long-necked birds.

The largest, most complex vertebrae are usually those of the thoracic or upper back region. These often bear long dorsal processes for the attachment of large muscles associated with the neck and shoulders, and transverse processes that serve as attachment points for the ribs. The lumbar vertebrae of the lower back are somewhat similar in form, but with more modest processes. The vertebrae of the pelvic region are known as sacral vertebrae. Those of tetrapods are fused into a rigid structure, the sacrum, which supports the pelvis. In birds the sacrum is fused fully with the other pelvic bones, forming one large, very strong structure called the synsacrum.

Features for flight

Flight places a unique set of stresses and strains on the thoracic region. The sternum of flying birds is greatly enlarged and bears a pronounced keel to provide attachment for the enormous breast muscles required to power flight. Flightless birds do not have a large sternum. The enlarged coracoid bone of flying birds offers some support, and the ribs are reinforced with special cross supports called uncinate struts. These resist compression of the rib cage, allowing the birds to continue breathing deeply during flight.

The tail

A tail is another characteristic feature of vertebrates. The tail is a continuation of the vertebral column in which the vertebrae are smaller and more simple than elsewhere. The spinal cord does not continue into the tail, so the caudal (tail) vertebrae lack a neural arch. In great apes, such as chimpanzees and humans, the tail is greatly reduced. It contains just four vertebrae, fused to form a bony bump—the coccyx—at the base of the spine. In birds, the tail is also reduced. The caudal vertebra are fused to form a structure called the pygostyle. This feature is an attachment point for the muscles that move the tail feathers, which play a vital role in controlling the aerodynamics of flight. In amphibians the tail vertebrae are fused to form the urostyle, a rod of bone that helps reinforce the pelvis.

Each of the 7 cervical vertebrae is long, giving the giraffe extra height to reach leaves high above the ground.

14 thoracic vertebrae

5 lumbar vertebrae

4 sacral vertebrae

20 tail vertebrae

◀ Giraffe
A giraffe's spine is made up of 50 vertebrae. Those in the neck (the cervical vertebrae) are unusually long, giving the animal the height necessary to reach leaves that are inaccessible to other terrestrial mammals.

Vertebrate appendicular skeleton

CONNECTIONS

COMPARE the lobe fin of a **COELACANTH** with the ray fin of a **TROUT**. It is easy to see which has more potential as a weight-bearing structure.

COMPARE the hind limbs of a **BULLFROG** with those of a **KANGAROO**. Both are long and powerful as an adaptation for a hopping mode of locomotion.

COMPARE the forelimb of the **PENGUIN** with that of a **DOLPHIN**. In both, the forelimbs function as flippers.

The term *appendicular skeleton* refers to the appendages (limbs and fins) and the girdles that attach them to the axial skeleton. In most vertebrates, there are two sets of paired appendages: pectoral appendages attached to a pectoral girdle at the front end of the animal and pelvic appendages attached to the pelvic girdle toward the back.

The girdles provide stability for the fins of fish, and in tetrapods they allow the weight of the body to be transferred through the limbs. The bones of the pectoral girdle are the scapulae or shoulder blades, the clavicles or collarbones, and the coracoid. In birds, the two clavicles are fused to make a single more rigid structure called the furcula or wishbone. The vertebrate pelvic girdle is made up of the ilium, ischium, and pubis, which may be separate or fused into one structure, the pelvis. The pelvic girdle of fish is a small, isolated structure embedded in soft tissue. That of tetrapods is much larger and is connected to the vertebral column in the region of the sacral vertebrae.

Fins

The paired fins of fish fall into two main skeletal types. By far the greater number of modern fish belong to the ray-finned group, called the Actinopterygii. Their pectoral and pelvic fins have no real skeletal support; instead, these fins have just a short stub of bones at the base. The fins themselves are webs of skin supported mainly by dermal rays, spokelike rods derived from the skin tissues. The more rigid fins of modern sharks and their relatives are supported by similar rods of cartilage. In contrast, the so called lobe-finned fish have robust, fleshy paired fins, supported by a much more substantial bony skeleton. There are many extinct groups of lobe-finned fish, including the Sacroptergii, which are probably the ancestors of tetrapods. Modern lobe-finned fish are now much less common and include lungfish and coelacanths.

The pentadactyl limb

The limbs of tetrapods (amphibians, reptiles, birds, and mammals) come in all shapes and

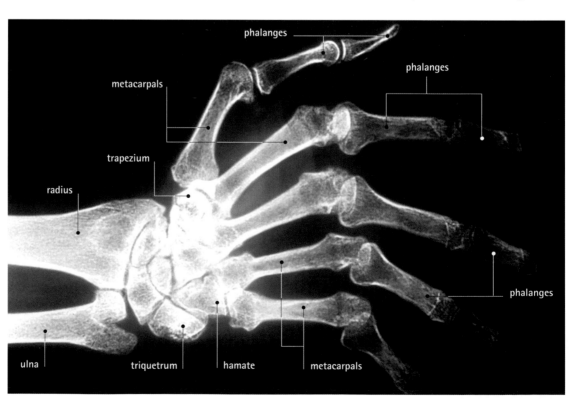

▶ *The five-fingered (pentadactyl) human hand has evolved from the pentadactyl plan of early amphibian ancestors. The hamate, the triquetrum, the trapezium, and several other small bones in the wrist are collectively called the carpals.*

phalanges

phalanges

metacarpals

phalanges

trapezium

radius

phalanges

ulna

triquetrum

hamate

metacarpals

Rhinoceros

The femur, tibia, and fibula are very robust and so able to support the heavy animal's weight. The rhino has an unguligrade stance: it walks on the tips of its toes.

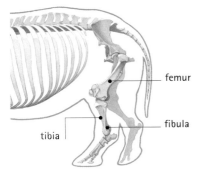

Hyena

The femur, tibia, and fibula are long and slender, and these bones are the most vulnerable of a hyena's skeleton. The stance is digitigrade: walking on the toes.

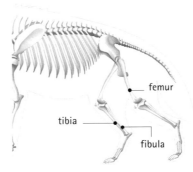

▼ VESTIGIAL HIND LEG AND PELVIS
Green anaconda

Structures that no longer have any function are called vestigial. Green anacondas and other boas have vestigial hind legs and a pelvis, which indicate that their ancestors had limbs.

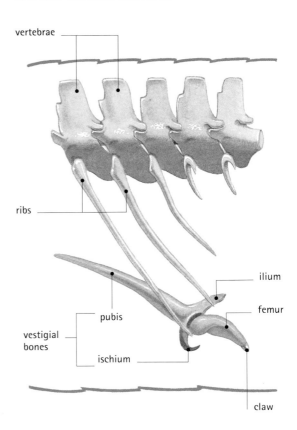

Ostrich

The femur is short, but the lower leg bones—the tibiotarsus and tarsometatarsus—are very long and ideally suited for fast running.

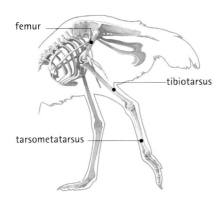

Kangaroo

The bones of the hind legs are robust, but whereas the femur is short, the tibia, fibula, and metatarsals are very long. This arrangement helps the kangaroo to leap.

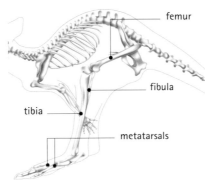

sizes, from the flippers and flukes of seals and whales to the wings of bats, the slender hooved legs of horses and antelope, and the dexterous hands and feet of the great apes. However, from a skeletal point of view, all these structures are similar. They are nearly all based on the five-fingered pentadactyl limb inherited from an early amphibian ancestor. The pentadactyl limb contains a basic set of bones, which vary in size and structure to create the startling variety of forms described above.

Bats and birds

The versatility of the pentadactyl limb is demonstrated to great effect in the forelimbs of flying vertebrates. The five-fingered plan is obvious in the wing of a bat, which is supported by a set of greatly elongated hand bones. They support the membrane of the wing in much the same way as the ribs of an umbrella—

▲ HIND LIMB BONES
The remarkable diversity in the arrangement of vertebrates' hind bones reflects extremely varied lifestyles.

opening out to hold the wing taut and closing to fold the whole wing away. Bat wing bones are extremely thin, and this characteristic makes them light but not very strong.

The pentadactyl plan is less easily recognized in the wing of a bird, where the bones of the wrist, hand, and fingers are reduced in number and modified in size and shape. The wing bones of birds are larger than those of bats; but being hollow, they are also very light. Their role is mainly supportive. Unlike bats, whose skeleton dictates the precise shape of the wing, birds fine-tune the shape their wings by adjusting the angles of the feathers.

AMY-JANE BEER

FURTHER READING AND RESEARCH
Raven, Peter H., George B. Johnson, Susan R. Singer, and Jonathan B. Losos. 2004. *Biology.* McGraw-Hill Science: New York.

Sloth

CLASS: Mammalia ORDER: Xenarthra
FAMILIES: Bradypodidae and Megalonychidae
GENERA: *Bradypus* and *Choloepus*

All five living species of sloths live in the tropical forests of Central and South America. There are three species of three-toed sloths, which have three toes on all their limbs; and two species of two-toed sloths, with two toes on their forelimbs but three toes on their hind limbs. All sloths hang upside down from branches, have a shaggy coat of fur, and have an extremely slow pace of life.

Anatomy and taxonomy
Scientists place all organisms into taxonomic groups based largely on anatomical features. Sloths are mammals that belong to an order called Xenarthra, along with the anteaters and armadillos. Sloths were once placed in the order Edentata, but this name is no longer used.

● **Animals** These organisms are multicellular and eat other organisms for energy. Animals differ from other multicellular life-forms in their ability to move from one place to another. Most animals have a nervous system that allows them to react rapidly to touch, light, and other stimuli.

● **Chordates** At some time in its life cycle, a chordate has a stiff, dorsal (back) supporting rod called the notochord that runs along most of the length of the body.

● **Vertebrates** The vertebrate notochord develops into a backbone (spine, or vertebral column) made up of units called vertebrae. The vertebrate muscular system that moves the head, trunk, and limbs consists primarily of muscles that are arranged in mirror image on either side of the backbone.

● **Mammals** Mammals are endothermic (warm-blooded) vertebrates with fur or hair made of keratin. Females have

▼ *There are five living species of sloths, in two families. The sloths' closest relatives are the anteaters and armadillos; they and the sloths make up the order Xenarthra. Only living species of sloths are shown in this family tree.*

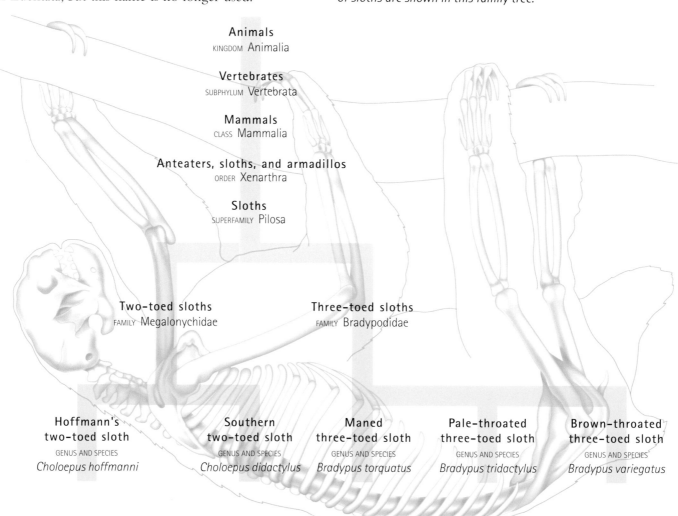

Animals
KINGDOM Animalia

Vertebrates
SUBPHYLUM Vertebrata

Mammals
CLASS Mammalia

Anteaters, sloths, and armadillos
ORDER Xenarthra

Sloths
SUPERFAMILY Pilosa

Two-toed sloths
FAMILY Megalonychidae

Three-toed sloths
FAMILY Bradypodidae

Hoffmann's two-toed sloth
GENUS AND SPECIES
Choloepus hoffmanni

Southern two-toed sloth
GENUS AND SPECIES
Choloepus didactylus

Maned three-toed sloth
GENUS AND SPECIES
Bradypus torquatus

Pale-throated three-toed sloth
GENUS AND SPECIES
Bradypus tridactylus

Brown-throated three-toed sloth
GENUS AND SPECIES
Bradypus variegatus

mammary glands that produce milk to feed their young. In mammals, the lower jaw is formed by a single bone. Mammalian red blood cells, when mature, lack a nucleus; the red blood cells of all other vertebrates have a nucleus.

● **Placental mammals** Also called eutherians, these animals nourish their unborn young through a placenta, a temporary organ that forms in the mother's uterus during pregnancy.

● **Xenarthrans** Xenarthrans are a mainly South American group, with a few species living in North America, where there were many more species in the recent past. All living xenarthrans lack incisors and canine teeth, but only anteaters are completely toothless. In other xenarthrans, the remaining teeth are simple and peglike and lack the covering of enamel that occurs on the teeth of other mammals. The number of cervical (neck) vertebrae varies from six to nine, depending on the species. This variation is very unusual: mammals, even giraffes, usually have seven neck vertebrae.

● **Armadillos** Armadillos are distinguished from other xenarthrans by their "shell." Covering most of the upper surface of the body, the shell is composed of a series of bony plates covered with a thin layer of horn. Flexible skin connects the plates; this allows some species to roll into a ball when threatened. The limbs and the top of the head are

▲ *The pale-throated three-toed sloth lives in the northern forests of South America. It is the most common three-toed sloth.*

also covered by the plates, and the tail is protected by rings of bone. An armadillo's strong pelvis and stout limb bones enable it to dig rapidly to make burrows or find food.

● **Anteaters** These xenarthrans are specialist eaters of ants. Anteaters have short, powerful forelimbs with long, sharp claws that are used to break into insect nests. A long, tapered snout and a long, sticky tongue help anteaters to reach inside nests. Anteaters have no teeth.

● **Sloths** There are two families of sloths: the three-toed sloths and the two-toed sloths. The number of toes refers to the front legs only; the hind limbs always have three toes. Sloths are slow-moving climbers that feed on leaves in trees. They descend to the ground roughly once a week to defecate. They have long limbs, with the forelimbs longer than the hind limbs. Two-toed sloths are covered with pale gray-brown fur that has a green hue because cyanobacteria grow on it. Although sloths are endothermic, their body temperature fluctuates enormously. Three-toed sloths even regulate their temperature by basking in the sun.

FEATURED SYSTEMS

EXTERNAL ANATOMY Sloths are four-legged, furry mammals suited to hanging upside down from branches by their strong claws. *See pages 1178–1183.*

SKELETAL SYSTEM Sloths' bones are light and slender and built to resist stress from tension rather than stress from compression or bending. *See pages 1184–1185.*

MUSCULAR SYSTEM Evolution has reduced sloths' muscles to the bare minimum necessary for climbing. *See page 1186.*

NERVOUS SYSTEM A sloth has a small brain relative to its body size. Its most important sense is smell. *See page 1187.*

CIRCULATORY AND RESPIRATORY SYSTEMS The blood carries oxygen from the lungs to the cells, where it is used to turn blood sugar into energy. *See pages 1188–1189.*

DIGESTIVE AND EXCRETORY SYSTEMS A sloth has a large, complex stomach containing symbiotic bacteria that turn plant fiber into useful food. *See pages 1190–1191.*

REPRODUCTIVE SYSTEM Sloths mate in trees. After a long gestation, females give birth to a single small but well-developed baby. *See pages 1992–1993.*

External anatomy

COMPARE the tree-climbing features of a sloth with those of a **SQUIRREL**. Both animals rely on their claws to support them, but a squirrel uses its claws in a completely different way, as bark-piercing crampons rather than grappling hooks.

Sloths are instantly recognizable by their upside-down, slow-motion lifestyle. They live in the trees of their native tropical American forests, where they feed almost exclusively on leaves. Their external anatomy is well suited for climbing slowly through the forest canopy, but sloths are not well adapted for life on the ground. They have difficulty standing upright, although they can swim well—an ability that is useful during seasonal flooding of the rain forests in regions such as the Amazon basin.

The **face** is flat, is covered by short hair, and has only limited muscular mobility.

The **eyes** are small for a mainly nocturnal animal and face forward to give binocular vision.

The **limbs** are long, especially the front limbs. They are equipped with stout claws that are strong enough to support the sloth's weight when hooked over branches.

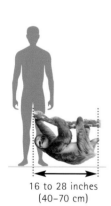

16 to 28 inches
(40–70 cm)

▶ Pale-throated three-toed sloth

This species of three-toed sloth has a rounded head with a flat face, small eyes, and inconspicuous ears. Its arms are long, and it has a stumpy tail. The coarse, shaggy fur coat is generally gray-brown but may appear green owing to cyanobacteria. The pale-throated sloth has a head-and-body length of about 16 to 28 inches (40–70 cm), the tail grows 0.8 to 3.5 inches (2–9 cm) long, and adults weigh 5 to 12 pounds (2.2–5.5 kg).

A sloth spends most of its life hanging upside down from branches. It feeds, sleeps, mates, and gives birth in this position and may even stay attached to a branch when it is dead. It spends much of its time asleep, especially during the day. It frequently hangs with all four feet bunched together on the branch, assuming a pear shape and looking more like a large, hairy fruit or a mass of leaves than an animal. This inconspicuous appearance may help it avoid the notice of predators. Sometimes, a sloth may rest in the fork of a branch or tree trunk, so its back is supported, but it never releases its grip on the branches and rarely turns the "right way up."

Toes and claws

A sloth owes its tenacious grip to the highly specialized structure of its feet. Their anatomical detail varies according to genus. All sloths have three functional toes on each hind foot, but the two-toed sloths of the genus *Choloepus* (Hoffman's sloth and Linne's sloth) have just

▲ SIZE COMPARISON
All the species of sloths differ little in overall size. A two-toed sloth (right) reaches a head-and-body length of 23 to 28 inches (58–71 cm), making it on average slightly bigger than its three toed relative (left).

*The **tail** is little more than a stump and has no real function.*

*The **body** is covered by a thick furry undercoat concealed by long guard hairs. These point downward when the sloth is hanging upside down, so they shed rain.*

Twos and threes

The toes of sloths are bound together within sheaths of muscle and skin and become separate only at the very tips, where they sprout long claws. In these illustrations of the front feet of three-toed and two-toed sloths, the claws are shown spread apart; they are normally held closer together.

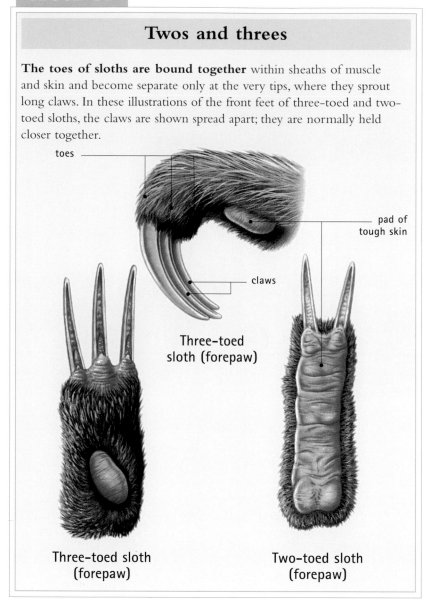

toes

pad of tough skin

claws

Three-toed sloth (forepaw)

Three-toed sloth (forepaw)

Two-toed sloth (forepaw)

two functional toes on each forefoot. The three-toed sloths of the genus *Bradypus* (pale-throated sloth, brown-throated sloth, and maned sloth) have three toes on each forefoot. The toes themselves are not visible, because their muscular parts are fused for their entire length. However, each digit ends in a conspicuous claw that extends from the fleshy part. There are two claws on the forelimbs of the two *Choloepus* species and three claws on the forelimbs of the three *Bradypus* species.

The curved claws are very long and sturdy, and by hooking them over horizontal branches, like coat hangers on a rail, a sloth can hang comfortably upside down. During active climbing, it clamps its claws against pads of tough skin covering its toes, giving a secure grip that enables it to climb vertical or near-vertical branches. It climbs hand over hand, with slowness and deliberation. Because a sloth rarely needs to move far to find its next meal, its inability to move fast is no problem. It can haul itself up large tree trunks when necessary—typically after its weekly visit to the ground to defecate—digging the tips of its claws into the bark to gain a purchase. The sloth's claws also make useful grappling hooks for dragging foliage within reach of its mouth during feeding.

Size and shape

Two-toed sloths are bulkier than three-toed sloths and some 25 percent heavier. Therefore, they have a smaller surface area in relation to their size, so they do not lose heat so easily and use less energy simply surviving. This factor may explain why two-toed sloths spend less time eating than their three-toed relatives.

Despite its greater bulk, a two-toed sloth has shorter limbs than a three-toed sloth. Also, the forelimbs and hind limbs of a two-toed sloth are of similar length, so the animal hangs beneath branches with its body roughly parallel to the branch. By contrast, the front limbs of a

Successful specialists

Modern sloths are very successful animals. In parts of tropical America, they account for more than a quarter of the total living weight—or biomass—of mammals; and on some islands they account for well over twice that. Yet they seem ill-qualified for success. Typical mammal success stories are species such as the brown rat, red fox, or common raccoon—all very adaptable animals that will eat a wide range of foods and live almost anywhere. Sloths, by contrast, are highly specialized and cannot survive outside their native forests. They owe their success to the great abundance of their primary food and their strange lifestyle, which ensures that they have no direct competitors.

COMPARATIVE ANATOMY

Armored relatives

The order Xenarthra, of which sloths are members, also contains two other groups: the anteaters and armadillos. The anteaters are superficially similar, but the armadillos look very different, with bodies that are covered with tough armor formed from horn-covered bony plates. The armor helps protect these ground-living animals from thorns and—especially in those species that can roll into an impregnable ball if attacked—from predators such as jaguars and pumas.

The pangolins—although not xenarthrans—are a group of African and Asian ant eaters. They have a similar type of armor. A pangolin's skin is protected by horny overlapping scales, resembling those of a fish. They are structurally very different from the bony plates of an armadillo and seem to have evolved separately as a response to similar environments—this is an example of convergent evolution.

▲ Yellow armadillo
Armadillos are native to the New World, ranging from South America to the southern United States. Their body is protected by tough horn-covered bony plates.

▶ Cape pangolin
There are seven species of pangolins living in Africa and Asia. Pangolins have overlapping horny scales to protect them from predators.

three-toed sloth are considerably longer than its hind limbs, so it tends to hang head downward from a horizontal branch. It is likely that this difference in limb lengths is a side effect of evolving, over millions of years, a longer arm reach to grasp more distant branches—something shared by many other tree-living mammals, such as gibbons. The other obvious difference between the two sloth genera is that three-toed sloths have a short, stumpy tail, whereas two-toed sloths have no tail.

Three-toed sloths have a small, round head with small ears, a short snout, and a broad, flat face with forward-facing eyes and a high forehead. The pale-throated and brown-throated sloths also have very short hair on their face. This combination of characteristics gives them a curiously appealing look, like a koala or even a human. Sloths' facial muscles are poorly developed, however, so sloths have very limited powers of expression. A two-toed sloth has a longer snout, giving a more bearlike appearance. This impression is accentuated when the animal opens its mouth to reveal four pointed teeth that resemble the long canines of a real bear.

EVOLUTION

Ancestral giants

Modern sloths are small, highly specialized climbers, but many of their ancestors were ground-living giants. Some species, such as the giant ground sloth *Megatherium,* were the size of elephants, growing to 20 feet (6 m) long and capable of rearing up on their hind legs to browse tree foliage. *Megatherium* belonged to one of three families of ground sloths that evolved some 40 million years ago on the then isolated continent of South America. There they had little competition from other types of mammals and flourished until the end of the last ice age, about 10,000 years ago. Some species may even have survived into historical times, because they appear in the legends of the native peoples of Patagonia.

▼ *Megatherium*
As big as a small elephant, Megatherium *was an herbivore. It walked on the sides of its hind feet and on the knuckles of its forefeet, as its relatives the anteaters do today.*

IN FOCUS

Winged lodgers

Since a sloth pays so little attention to grooming, it is easy for other organisms to live in its fur. These organisms include the bacteria that tint green the long outer guard hairs during the wet season, as well as a species of small moth. Up to 100 of these moths may live on a single sloth, and they were once thought to feed on the cyanobacteria. In fact, the adult moths do not eat the bacteria, but they lay their eggs in the sloth's dung, which the animal deposits at the foot of trees in habitual latrines. The eggs hatch into caterpillars that feed on the dung, pupate, and turn into adult moths. The moths then climb onto the sloth when it next visits the latrine, and they use the animal as transport to other latrines where they can lay their eggs.

Shaggy coat

All sloths have a thick pelt to help conserve heat and minimize energy loss. The short, fine underfur is protected by a shaggy coat of much longer, coarser guard hairs. These hairs keep the underfur dry by shedding water during tropical rainstorms, and to assist with this the long hairs typically hang downward. Thus the hair lies in the opposite direction from that in other mammals, pointing away from the animal's feet.

The fur of two-toed sloths is basically grayish brown but paler on the face and absent on the muzzle, where dark naked skin is revealed. Three-toed sloths are similar in color, but the different species have their own variations. The maned three-toed sloth has long, dark hair on its head and neck, with paler eye patches and muzzle. The pale-throated three-toed sloth has a marbled pattern of pale and darker brown fur on its body and a pale face with dark masklike markings; males have a darker patch of brown and black on the back. The brown-throated three-toed sloth is very similar, apart from its dark throat patch, but it is generally a more uniform pale brown.

All sloths tend to have a greenish tinge owing to the cyanobacteria that grow on their fur, especially during the wet season when the animals are always slightly damp. The green tinge helps camouflage them in the tree foliage, concealing them from aerial predators such as harpy eagles.

To some extent, the colonization of a sloth's fur by cyanobacteria is an accidental side effect of the animal's relative immobility. Because a sloth does not move fast, its fur is never subjected to friction, and—unlike a cat, for example—it does not groom its coat to keep the fur clean. However, the structure of the hair is also unique, with microscopic grooves running along the hair shaft that provide the cyanobacteria with a microhabitat.

The cyanobacteria—formerly known as blue-green algae—consist of microscopic single cells that produce sugars from carbon dioxide and water using the energy of light, in the process of photosynthesis. It has been suggested that the sloth benefits nutritionally from this by licking the bacteria or even by absorbing the sugars through its skin. This theory has yet to be proved, however, and the sloth may simply tolerate the cyanobacteria in the same way that it tolerates the small moths that often live in its fur.

▼ *Hoffman's sloth is the smaller of the two-toed sloths. Its range stretches from northern Brazil to Nicaragua in Central America.*

Skeletal system

CONNECTIONS

COMPARE the cheek teeth of a sloth with the very high-crowned cheek teeth of a *ZEBRA*. The sloth's teeth continue to grow as they are worn away by chewing tough leaves, so they never wear out. The zebra's teeth do not continue growing, but they are much bigger and have plenty of scope for wear.

The skeleton of most moderate-size to large land mammals serves to support their weight by propping them up, somewhat like the legs and frame of a chair. The main bones are typically thick and strong to resist compression and are made of collagen—a tough but springy protein—reinforced with hard calcium compounds. The calcium compounds provide most of the strength but are brittle; the protein gives the bone a degree of flexibility that keeping it from snapping under the normal stresses of life.

The limb bones of a sloth are made of the same material, but they are subjected to different stresses. Instead of supporting the animal's weight from below, they act as mobile suspension rods, and have to be strong under tension. They do not have to be thick to resist bending, so they

IN FOCUS

Bone cells

The bones of a mammal such as a sloth originate as cartilage, which consists mainly of the tough protein collagen. Each future bone is covered by a membrane containing special cells called osteoblasts, which produce a substance that makes the blood release calcium minerals. These impregnate the cartilage and harden it to form bone. This process is called ossification; it begins in the middle of long bones such as those of the legs and gradually extends toward the ends. The bone stops growing once it is fully ossified.

▼ **Pale-throated three-toed sloth**
The limb bones of sloths are long and slender with highly mobile wrist and ankle joints. Three-toed sloths have extra cervical vertebrae, allowing them to rotate their neck up to 270 degrees.

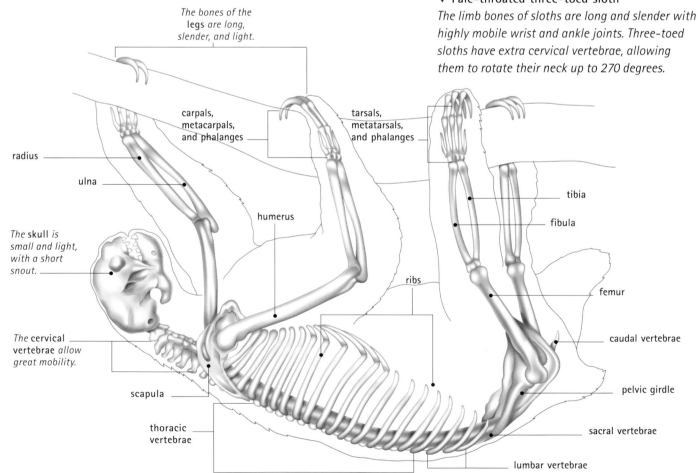

The bones of the legs are long, slender, and light.

carpals, metacarpals, and phalanges

tarsals, metatarsals, and phalanges

radius

ulna

tibia

fibula

humerus

The skull is small and light, with a short snout.

ribs

femur

The cervical vertebrae allow great mobility.

caudal vertebrae

scapula

pelvic girdle

thoracic vertebrae

sacral vertebrae

lumbar vertebrae

Two-toed sloth

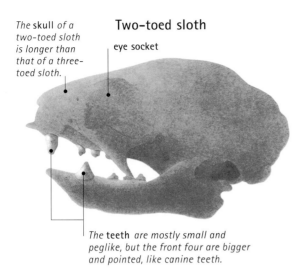

The skull of a two-toed sloth is longer than that of a three-toed sloth.

eye socket

The teeth are mostly small and peglike, but the front four are bigger and pointed, like canine teeth.

Three-toed sloth

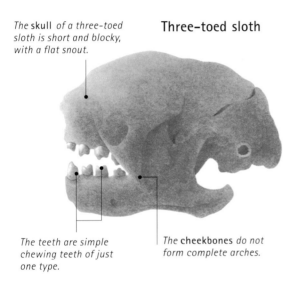

The skull of a three-toed sloth is short and blocky, with a flat snout.

The teeth are simple chewing teeth of just one type.

The cheekbones do not form complete arches.

▲ SKULLS

Two-toed sloths have a longer skull than their flat-faced three-toed cousins. In both types of sloths, the cheekbones are not complete.

are slender. The forelimb bones of a three-toed sloth in particular are also long, giving it extra reach as it clambers through the forest canopy.

The feet are also long and have functional toe bones that extend into the claws to give them essential reinforcement. The nonfunctional toes are reduced to slivers of bone. The toes are articulated to the limb bones by highly mobile wrist and ankle bones.

Mobile spine

The spine of a sloth has the same basic structure as the spine of any vertebrate. It is made up of a chain of separate bones called vertebrae and forms a strong column

Hand

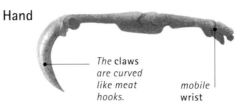

The claws are curved like meat hooks.

mobile wrist

enclosing the spinal cord. Sliding surfaces, or articulations, between the vertebrae allow the bones to move in relation to each other, giving the spine flexibility.

The sloths and their near relatives are unique among mammals in having extra articulations, called xenarthrales, in the lumbar region of the lower back. They have no obvious function in a sloth species, but they provide valuable spine reinforcement for armadillos during digging. Three-toed sloths have extra neck vertebrae that enhance neck mobility, allowing them to turn their head 270 degrees, like owls.

The skull of two-toed sloths is longer than that of the flat-faced three-toed sloths, but in both types the skull is relatively small and light. The teeth are also reduced compared with those of most mammals. A three-toed sloth has a row of peglike cheek teeth in each jaw—five in each side of the upper jaw and four in each side of the lower. They are all more or less the same, with cupped grinding surfaces for chewing tough leaves. The teeth do not have a hard enamel coating, but they grow continually to compensate for wear. There are no canines or incisors, but two-toed sloths have two large, pointed cheek teeth in each jaw that look very like the canines of other mammals.

Foot

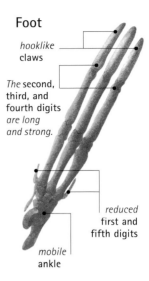

hooklike claws

The second, third, and fourth digits are long and strong.

reduced first and fifth digits

mobile ankle

▲ HAND AND FOOT

The hands and feet are suited for life in the trees. The outer digits of the hands and feet are greatly reduced. Both hands and feet have hooklike claws for gripping onto branches.

COMPARATIVE ANATOMY

Toothless anteaters

Sloths, armadillos, and anteaters were once classified together with pangolins and aardvarks in the order Edentata, a name that means "without teeth." In fact, both sloths and armadillos have teeth—a giant armadillo may have up to 100—but the anteaters and ant-eating pangolins are truly toothless. A pangolin, for example, does not need to chew, because the ants that it gathers with its tongue are ground to a pulp in its stomach. Its lower jaw is reduced to a bladelike structure with no real mobility or chewing muscles, and because the skull has no attachment points for such muscles it is little more than a simple cone.

Muscular system

COMPARE the slender build of a sloth with the massive muscles of one of its enemies, the **PUMA**. Power is vital to the puma's success as a predator, just as economy is vital to the sloth's success as a leaf eater.

The ground-living prehistoric ancestors of sloths probably had a typical mammalian muscular system, but as sloths became more suited to living in trees and eating leaves, their muscles gradually became more specialized for an arboreal (tree-living) life.

One of the most important factors in sloths' evolution has been their diet, which consists almost entirely of leaves that have relatively little nutritional value. This dietary restriction has favored individuals with reduced muscle mass, because they need to extract less energy from their food. Over millions of years, sloths' skeletal muscles—the muscles attached to the bones responsible for movement—have slimmed down to account for just 25 to 30 percent of their weight. This is roughly half the figure typical of land mammals and is one reason why sloths move so slowly. In engineering terms, they have a low power-to-weight ratio.

Drastic reductions

The sloth's most important muscles are those associated with climbing, so its limb muscles account for a high proportion of the animal's muscle mass. Yet limb muscles that in other mammals permit such a basic action as walking have been so drastically reduced that they barely function, and a sloth cannot walk upright. It has to drag itself along while sprawled on the ground, and it makes an easy target for predators until it finds a tree and climbs to safety.

Other nonessential muscles in the sloth have undergone the same process of reduction. For example, facial muscles—except those muscles that are used for chewing—are not important to a basically solitary animal that has no need to communicate using facial expressions. These muscles have been reduced to the minimum necessary for survival, and so a sloth has only one facial expression.

▶ **Pale-throated three-toed sloth**
Compared with other mammals, sloths are not muscular. The limb muscles, which are used for climbing, make up most of a sloth's muscle mass.

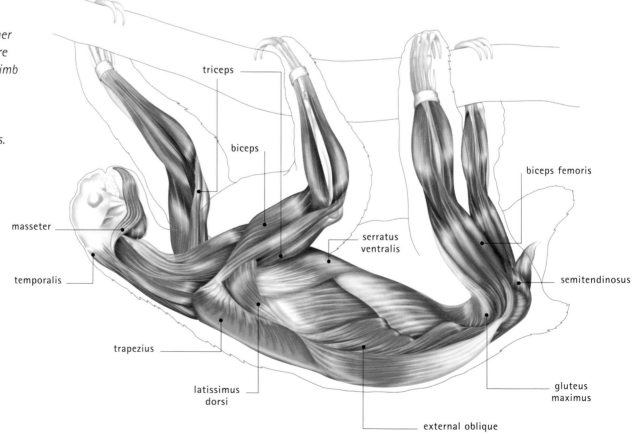

triceps

biceps

biceps femoris

masseter

serratus ventralis

temporalis

semitendinosus

trapezius

gluteus maximus

latissimus dorsi

external oblique

Nervous system

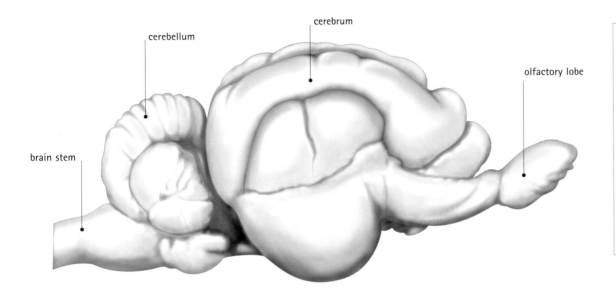

cerebellum

cerebrum

olfactory lobe

brain stem

COMPARE the binocular vision of a sloth with the vision of a *HARE*. The hare's eyes are mounted high on each side of its head, giving it a panoramic view of approaching danger, but it has a poor appreciation of distances.

CONNECTIONS

A sloth's actions are controlled by electrical nerve impulses that pass through an elaborate network of nerve fibers. Some of the fibers gather information from sensory cells and pass it to the brain and spinal cord. There, the information is processed, and appropriate signals are sent by way of the motor nerves to the muscles. Much of the processing of information is automatic, or virtually so, because a sloth does not need to give much thought to survival. As a leaf-eating animal, it is surrounded by food, and its camouflage provides a passive defense against predators.

Pungent secretions

A sloth has poor hearing, with small ears that are usually hidden by its fur. Its eyesight is relatively good, however, and able to discern colors. Both eyes face forward, providing the binocular vision that allows the animal to assess distances accurately. This may seem irrelevant to the life of a slow-moving herbivore, but it could be important in the forest canopy, where a misjudgment of the gap between branches might lead to a fatal fall.

A sloth's most important sense is olfaction, or smell, which enables it to locate suitable leaves to eat and also interact with other sloths. Males, in particular, are believed to scent-mark

▲ BRAIN
Linne's two-toed sloth
The sloth has large olfactory lobes because olfaction, or smell, is the animal's most important sense. It is used for finding food and communicating.

branches with pungent secretions from their anal glands. This behavior probably enables neighboring males to keep track of one another's movements within the forest and even identify individuals.

EVOLUTION

Colorful inheritance

Sloths are most active at night, and two-toed sloths are almost entirely nocturnal. Despite this, their eyes are small and sensitive to color. Typically, nocturnal animals have large eyes to gather as much light as possible, and most of the light-sensitive cells in each eye are of a type that react to dim light but cannot detect color. The vision of sloths seems more suitable for animals that are active by day, and this characteristic may be an inheritance from their giant, ground-living ancestors. Because sloths depend mainly on their sense of smell, their small, color-sensitive eyes have not decreased their ability to survive and breed. Therefore, there has been little pressure over millions of years for evolutionary processes to result in sloths with more light-sensitive eyes.

Circulatory and respiratory systems

COMPARE the slowed-down body functions of a sloth with the very much quicker functions of a *RAT*. The rat has a much faster metabolic rate, which is reflected in its actions. The sloth seems to operate in slow motion, but the rat looks as if it is running in fast-forward.

As a sloth digests leaves, simple sugars, such as glucose, that were made by the leaves are absorbed into its bloodstream. These sugars contain energy derived from the sun locked into their chemical structure during their synthesis by the plants. If the sugars are oxidized (effectively, burned), the energy is released. The sloth's body makes use of this chemical reaction to fuel its muscles and vital organs.

Oxidation requires a supply of oxygen, and the sloth gets this from the air that it draws into its lungs when breathing. Oxygen from the air passes through the thin walls of tiny air sacs in the lungs and into the blood that is pumped through the lungs by the heart. The oxygen-rich blood then returns to the heart, which pumps it through the arterial system to every living cell of the sloth's body.

The blood carries both sugar and oxygen, but the two react together to generate energy only within the target cells in a process called

IN FOCUS

Energy cells

The oxygen and blood sugar (glucose) carried in a sloth's blood (and that of other mammals) are converted into energy inside the cells of its muscles and other living tissues. Each cell contains mitochondria. They are the sites of respiration: the "slow burn" that oxidizes glucose to liberate its stored energy. Respiration turns glucose and oxygen into carbon dioxide and water, reversing the process of plant photosynthesis, which uses carbon dioxide and water to make glucose and oxygen. The photosynthesis reaction absorbs energy from the sun, and when the reaction is reversed within the sloth's cells by respiration, the energy is released.

▶ **Pale-throated three-toed sloth**
Like other mammals, sloths have a double circulation. The four-chamber heart pumps blood to the lungs to pick up oxygen. This blood then returns to the heart to be pumped around the body, supplying cells and tissues with oxygen for respiration. On average, a three-toed sloth's heart beats about 78 times a minute, a little faster than the human heart.

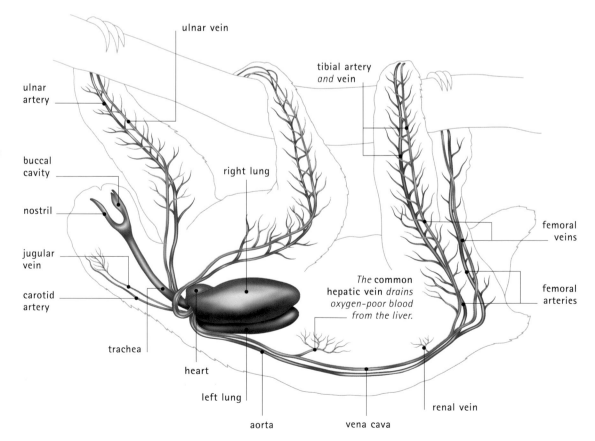

ulnar vein

tibial artery *and* vein

ulnar artery

buccal cavity

right lung

nostril

jugular vein

carotid artery

femoral veins

The common hepatic vein drains oxygen-poor blood from the liver.

femoral arteries

trachea

heart

left lung

renal vein

aorta

vena cava

respiration. In this process, the sugar and oxygen are converted into water and carbon dioxide. These waste products pass into the blood, which flows back to the heart through the animal's network of veins. The heart then pumps the oxygen-depleted blood back to the lungs, where the water and carbon dioxide pass out and fresh oxygen passes in.

Slow turnover

The conversion of sugar and oxygen into energy is one of the many chemical reactions that occur in a sloth's body. Together, these reactions are called metabolism. Some animals, such as shrews, have a very fast metabolism, and this speeds up everything they do. Sloths, on the other hand, have a very slow metabolism, which slows everything they do. Their muscles and organs turn food into energy at less than half the rate typical for animals of their weight. They move very slowly, digest their food very slowly, and react very slowly.

The slow metabolism of sloths helps them cope with their leafy diet, which has a very low nutritional value in relation to its bulk. If a sloth's body chemistry operated at a normal rate, it would get through more energy than it could replace by eating. Clearly, such a low metabolic rate is partly counterproductive, because it reduces the rate at which the sloth can gather and digest food. The animal never has any trouble filling its stomach, however, so on balance its slow metabolism serves it well.

▲ *A brown-throated three-toed sloth climbs a tree. Sloths have a slow metabolism and do everything, including climbing, very slowly.*

IN FOCUS

Unstable temperature

Most mammals maintain a steady body temperature, regardless of the temperature of their surroundings. This allows them to keep active even in freezing conditions. The body heat is generated by the conversion of food into energy, but the leafy diet of a sloth is so low in nutritional value and its metabolism so slow that it cannot obtain enough energy to maintain a constant temperature. Its muscles also operate at such a reduced rate that it cannot keep warm by shivering. Its body temperature is never more than 9°F (5°C) above the temperature of its environment and often fluctuates by about 18°F (10°C) daily, falling as low as 86°F (30°C) at night. If it cooled down any more, it might die, so sloths cannot live outside the tropical zone.

Digestive and excretory systems

COMPARE the complex digestive system of a sloth with the equally complex but very different system of a *WILDEBEEST*. Both animals eat leafy plant food, but they have evolved different ways of digesting it. Both also rely on symbiotic bacteria to break down the cellulose in their diet.

Many animals eat plants, but most select the most nutritious, easily digested parts, such as juicy young shoots, flowers, fruits, and energy-storing roots. Some plant-eating animals, however, concentrate on eating leaves. These are easier to find but much harder to digest because their cells have tough walls of woody plant fiber, or cellulose. They have relatively little nutrient value, so the animals need to eat a much greater bulk of food to get the nutrients they need.

Sloths are among the most specialized of these leaf eaters. They have evolved a digestive system capable of eating a lot of food in a short time and processing it over a long period. The main digestive organ is the very large, complex stomach, which, when full, accounts for almost a third of a sloth's weight. It is divided into five main chambers, with three on the left and two on the right.

IN FOCUS

Picky feeder

In a study of maned sloths—the rare three-toed species living only in the Atlantic coastal forests of Brazil—the animals were found to eat roughly 70 percent young leaves, which contain fewer toxins than mature foliage. Each sloth also selected leaves from just a few types of trees, which varied from one individual sloth to another. This suggests that each sloth develops a resistance to a particular cocktail of plant poisons, and is aware of it. This variation between individuals may also help reduce competition between neighboring sloths, enabling more sloths to feed in the same patch of forest.

▶ **Pale-throated three-toed sloth**
To cope with its diet of leaves, the sloth has a complex five-chamber stomach that grinds leaves and contains bacteria that break down the tough carbohydrate cellulose, which is the main constituent of plant cell walls.

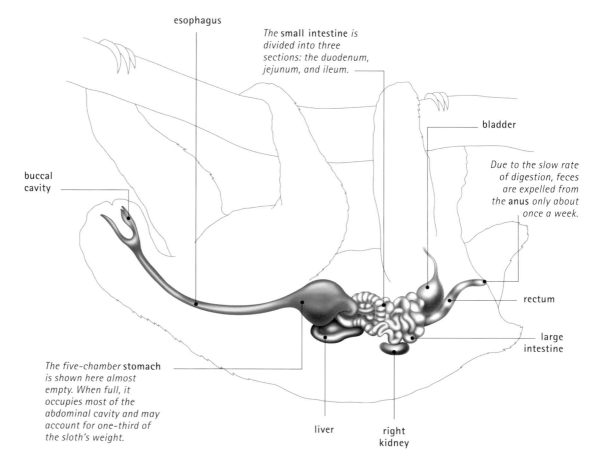

esophagus

The small intestine is divided into three sections: the duodenum, jejunum, and ileum.

bladder

Due to the slow rate of digestion, feces are expelled from the anus only about once a week.

buccal cavity

rectum

large intestine

The five-chamber stomach is shown here almost empty. When full, it occupies most of the abdominal cavity and may account for one-third of the sloth's weight.

liver

right kidney

The three left-hand chambers have a tough, horny lining that helps grind the chewed leaves to a pulp, crushing the plant cells to release their juices. The food then passes to the right-hand chambers, where it is mixed with enzymes secreted by the stomach. The enzymes break down the starches, proteins, and fats into sugars and other substances that can be absorbed into the sloth's bloodstream. The digested food then passes through the relatively short intestine, where more nutrients and water are absorbed. The whole process takes about a month, partly because the sloth's low metabolic rate slows the rate of digestion.

Symbiotic bacteria

The stomach chambers contain bacteria that are able to break down cellulose—a type of carbohydrate, like starch—into sugar. A mammal's digestive enzymes cannot do this, and cellulose passes through the digestive systems of many animals virtually undigested. The symbiotic relationship with bacteria is therefore vitally important to plant eaters and occurs in many mammals as diverse as cattle, porcupines, and koalas.

The sloth's digestive system is also able to neutralize toxic chemicals such as alkaloids and tannins that occur in the leaves of forest trees. These chemicals defend plants' leaves against leaf-eating organisms, and the chemicals certainly deter many animals. Even sloths prefer to eat the less toxic younger foliage, yet despite this their teeth are stained brown by the high levels of tannins in their diet. The neutralization process is probably helped by the long period of digestion in the stomach.

A sloth rarely needs to defecate, owing to its slow digestive process. It does so just once a week, using a habitual latrine at the foot of a tree. Perhaps surprisingly, there is still a lot of food value left in the dung, and it is recycled by at least nine species of forest insects.

COMPARATIVE ANATOMY

Stomach for the job

Although sloths and anteaters are classified in the same order, Xenarthra, their digestive systems are very different to cope with their specialized diets. The digestive system of a leaf-eating sloth is much more complex, with a large, multichamber stomach containing cellulose-splitting bacteria for converting plant fiber into sugar. An anteater does not need to do this, and its stomach is smaller and simpler. Even its secretions are simplified, for unlike a typical mammal it does not produce hydrochloric acid to help with digestion. Instead, an anteater relies on the formic acid in its ant prey to break down their tissues.

In one respect, however, sloths and anteaters face the same problem. The most nutritious parts of both mature plant cells and ants are encased in tough capsules of either cellulose or chitin—the horny material that forms the exoskeleton of an insect. These capsules must be broken open if the animal is to get the most from its food, so both sloths and anteaters have a tough, muscular stomach lining that mashes their food and grinds it to a pulp.

Reproductive system

COMPARE the small but well-developed baby of a sloth with the tiny newborn cub of a *GRIZZLY BEAR*. The bear cub weighs less than 1 percent of an adult bear and needs large quantities of milk if it is to grow healthily. This demand places a big strain on the mother, who loses a lot of weight as a result—something that a sloth could not survive.

Male and female sloths look identical. Even on close examination they are hard to tell apart, because the male does not have prominent external genitalia like those of many other mammals. The penis is very small and barely visible, and the testes lie inside the animal's body instead of in an external sac, or scrotum. The body temperature of most mammals is too high for male sperm to develop properly, so the testes are usually external. However, the typically low body temperature of a sloth makes this unnecessary. A female sloth also lacks conspicuous sexual characteristics, having just a single genito-urinary duct instead of two separate ducts. She has two milk-producing teats, but these are normally hidden by the thick fur on her chest.

The physical similarity of male and female sloths is typical of animals that have a solitary life and that do not engage in elaborate courtship rituals or social behaviors that favor large or flamboyant males. Courtship is perfunctory, but it is possible that females select stronger, healthier males on the basis of the scent signals that males leave on forest trees. The pairs mate, face to face, in a tree. They

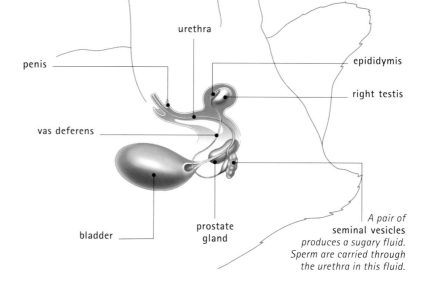

▶ MALE REPRODUCTIVE SYSTEM
Male sloths have internal testes, which produce sperm, and a relatively small penis.

urethra
penis
epididymis
right testis
vas deferens
bladder
prostate gland
A pair of seminal vesicles *produces a sugary fluid. Sperm are carried through the urethra in this fluid.*

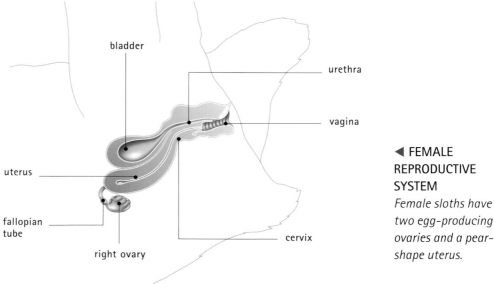

bladder
urethra
vagina
uterus
fallopian tube
right ovary
cervix

◀ FEMALE REPRODUCTIVE SYSTEM
Female sloths have two egg-producing ovaries and a pear-shape uterus.

Marsupial cousins

The uterus of a female sloth has a divided form that has much in common with the double uterus of a marsupial such as an opossum. The uteruses of the other xenarthrans—anteaters and armadillos—are the same. This fact suggests that the placental xenarthrans and early marsupials may have evolved from the same ancestor, whereas other placental mammals had different origins.

where the seasons vary little—so scientists are unable to explain its extended pregnancy.

Small but perfect

Sloths and their relatives are placental mammals, which keep babies in the uterus as they develop. The unborn young are nourished through the umbilical cord, which is attached to an organ called the placenta. This functions somewhat like a plug implanted in the wall of the uterus, gathering foods and other vital substances from the mother's bloodstream.

The single baby is small when born, weighing just 5 percent of its adult weight; yet it is very well developed in other ways. It is covered with fur and is able to see, hear, and cling to its mother. It is born headfirst while its mother hangs from a branch; it then immediately climbs up onto its mother's breast and begins to suckle. It feeds on milk for just four weeks but is carried by its mother for up to nine months and eats the same limited variety of leaves. The newborn may need a long gestation to enable it to be sufficiently well developed at birth to survive in the trees. However, its low birth weight and short suckling period reduce the drain on its mother's limited nutritional resources.

JOHN WOODWARD

separate soon after, and males probably mate with other females if they get the opportunity.

In the north of their range, pale-throated three-toed sloths breed seasonally, so the young are born between July and September, after the rainy season. Otherwise, sloths seem to breed at any time of year. The gestation period for most species is relatively long—six months—but Hoffman's two-toed sloth appears to have an even longer gestation of between 11 and 12 months. This long period could be caused by delayed implantation of the fertilized egg in the wall of the uterus. Delayed implantation is typical of mammals that live in very seasonal climates and benefit from delaying birth until a suitable time of year. However, this does not apply to Hoffman's two-toed sloth, which lives from Nicaragua to central Brazil—regions

▲ *A baby sloth is born small but relatively advanced—it has a furry coat and can see, hear, and cling to its mother. It stays with its mother for about nine months.*

FURTHER READING AND RESEARCH

Macdonald, David. 2006. *The Encyclopedia of Mammals.* Facts On File: New York.

Nowak, R. M. 1999. *Walker's Mammals of the World* (6th ed.). Johns Hopkins University Press: Baltimore, MD.

Snapping turtle

ORDER: Chelonia, or Testudinata SUBORDER: Cryptodira
FAMILY: Chelydridae

The two or three species of snapping turtles have a large head and powerful jaws. Alligator snapping turtles live in rivers, lakes, and swamps in the southern parts of the United States. They are among the biggest freshwater turtles in the world. Common snapping turtles are much smaller. They, too, live in North America, but they have a much wider distribution than alligator snappers. The big-headed turtle lives in rocky streams in parts of southeastern Asia. There is disagreement among herpetologists (scientists who study reptiles and amphibians) whether the big-headed turtle should be placed in the same family as snapping turtles.

Anatomy and taxonomy
Turtles and tortoises are reptiles, a group that also includes lizards, snakes, crocodilians, worm lizards, and tuataras. All turtles and tortoises have an outer shell, although in a few species this is covered externally by soft skin. Turtles and tortoises are ancient groups of animals. Fossils that are recognizably turtles have been found from the Triassic period, about 200 million years ago.

● **Animals** All animals are multicellular (many-celled) and depend on other organisms for food. They differ from other multicellular life-forms in their ability to move around and in their rapid response to stimuli.

● **Chordates** At some stage in its life a chordate has a stiff dorsal (back) supporting rod called a notochord.

● **Vertebrates** In vertebrates, the notochord develops into a backbone, or spine, made up of units called vertebrae. Vertebrates have a system of paired muscles lying on either side of the line of symmetry.

▼ *Herpetologists divide living turtles and tortoises into 12 families: 10 families of hidden-necked turtles and 2 families of side-necked turtles. On this family tree, the snapping turtles and the big-headed turtle have been placed in separate families, but some scientists believe they should be in the same family.*

Animals
KINGDOM Animalia

Chordates
PHYLUM Chordata

Vertebrates
SUBPHYLUM Vertebrata

Reptiles
CLASS Reptilia

Snakes and lizards
ORDER Squamata

Turtles
ORDER Chelonia,
or Testudinata

Crocodilians
ORDER Crocodylia

Side-necked turtles
SUBORDER Pleurodira

Hidden-necked turtles
SUBORDER Cryptodira

Softshell turtles
FAMILY Carettochelyidae

Snapping turtles
FAMILY Chelydridae

Big-headed turtle
FAMILY Platysternidae

Sea turtles
FAMILY Cheloniidae

Tortoises
FAMILY Testudinidae

● **Reptiles** Reptiles are four-legged vertebrates (although some have lost their legs during the course of evolution) and have a thick, horny skin, usually divided into plates called scales. Most reptiles lay eggs, but in some species the eggs are retained within the body of the female until they are ready to hatch. Most reptiles are unable to generate their own body heat as birds and mammals do. They can, however, regulate their body temperature by going into hot places when they need to warm up and cooler places when they need to lose heat.

● **Turtles and tortoises** There are approximately 250 living species, and they make up the order Chelonia (sometimes called Testudinata). Collectively, the turtles and tortoises are usually called chelonians. All of them have flattened ribs that support a shell. Usually the shell is made of bony plates on the inside and scales of hardened skin, called scutes, on the outside. In a few turtles, the scutes are softer and have a leathery texture. Another characteristic feature of turtles and tortoises is that they have no teeth.

● **Hidden-necked turtles** In these turtles, the head can usually be withdrawn into the shell for protection. This is achieved by bending the neck in a vertical (up-and-down) plane. Many species of hidden-necked turtles live in freshwater; some live on land; and eight species live in the sea. There is another, smaller group of turtles in which the neck is bent sideways. They cannot withdraw their head into the shell. These are called side-necked turtles. They are found only in freshwater and are confined to South America, Africa, and a few adjacent islands in the Indian Ocean, Australia, and New Guinea.

● **Snapping turtles** These are hidden-necked turtles with a very large head and powerful jaw muscles. The lower part of the shell (called the plastron) is smaller than in most

▲ *An alligator snapping turtle is able to open its jaws very wide, revealing the pink wormlike lure on its tongue.*

other turtles. Their tail is longer than that of most other chelonians. The tail of the common snapping turtle is toothed along its dorsal (top) surface.

FEATURED SYSTEMS

EXTERNAL ANATOMY Like all chelonians, snapping turtles have a shell. Compared with other species they have a large head and powerful jaws. *See pages 1196–1198.*

SKELETAL SYSTEM The shell is supported by flattened ribs, which, together with a number of other flattened bones, form the inner part of the shell. The outer part of the shell is made of plates of hardened skin. *See pages 1199–1200.*

MUSCULAR SYSTEM The body muscles are relatively poorly developed, but the muscles of the limbs and neck are powerful. *See page 1201.*

NERVOUS SYSTEM Migrating turtles can sense Earth's magnetic field, but scientists do not know exactly how this is achieved. *See page 1202.*

CIRCULATORY AND RESPIRATORY SYSTEMS In addition to using lungs, many aquatic turtles—including snapping turtles—obtain oxygen from their skin, from the mouth cavity, and from two anal sacs at the posterior end of the alimentary canal. *See pages 1203–1204.*

DIGESTIVE AND EXCRETORY SYSTEMS Snapping turtles and other turtles that feed mostly on animals have a shorter digestive tract than those species of turtles that feed mostly on plant material. *See page 1205.*

REPRODUCTIVE SYSTEM All female turtles lay eggs with shells, and all female turtles come out of water and onto land to lay their eggs. Most species bury their eggs. *See pages 1206–1207.*

External anatomy

CONNECTIONS

COMPARE the scutes of a snapping turtle with the scales of a *CROCODILE*. The scutes and scales serve to protect these reptiles; are ridged, or keeled, on the animals' upper surface; and are underlaid by bony plates.

Snapping turtles are large freshwater species, and alligator snapping turtles are giants among the freshwater turtles. Big males may have a shell that is more than 31 inches (80 cm) long. The biggest alligator snapper ever recorded weighed 251 pounds (143.3 kg), but most specimens are much smaller; females are much smaller than males.

Alligator snappers live in deep, slow-moving rivers, lakes, and swamps. They live in the Mississippi–Missouri basin, from eastern Texas and northwestern Florida northward to Illinois and Iowa. They are slow-moving animals and rarely come out of the water, although females come onto land to lay their eggs, and males occasionally emerge from the water to bask in the sunshine. Snapping turtles are dull brown in color: the skin is often a paler shade of brown than the shell. In older animals, the shell may become covered with algae.

Common snapping turtles wander farther from water than alligator snappers, occasionally

The tongue has a pinkish, wormlike **lure**, which the turtle can move. This action attracts prey to the turtle's open mouth when the turtle is submerged.

keel

▶ **Alligator snapping turtle**
The turtle is named for its rough, keeled carapace and scaly tail. These features have a texture like that of an alligator's back.

The head has a hooked **beak**. The head is large and well armored with tough, bony plates called scutes.

The **forelegs** are too large to be drawn into the turtle's shell. Each leg has five clawed digits.

31 in (80 cm)

claws

► **Alligator snapping turtle**
The turtle has a very large head, so even with the neck drawn under its carapace as far as it will go, the head is still not hidden. Strong scales with bony plates combine with the strong jaws and sharp beak to protect the head.

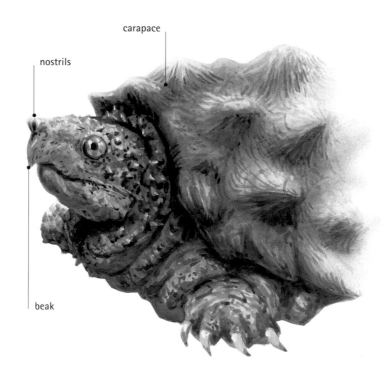

venturing up to 1.5 miles (2.4 km) away. They do not grow to the size of an adult alligator snapping turtle, and the average adult weighs about 30 pounds (14 kg), although specimens weighing 70 pounds (32 kg) have been recorded. The carapace is typically 16 inches (40 cm) long. This species has an extensive distribution: it lives in southern Canada, the United States east of the Rocky Mountains, parts of Central America, and western Colombia.

Alligator snappers have an enormous head, which cannot be retracted into the front part of the shell. The jaws have no teeth. Both the

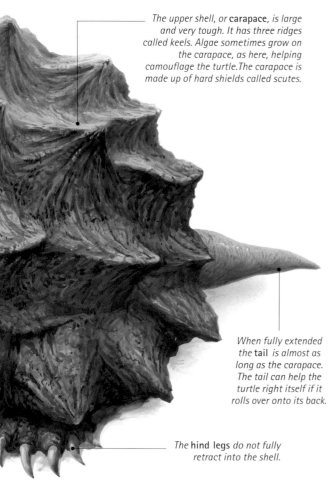

*The upper shell, or **carapace**, is large and very tough. It has three ridges called keels. Algae sometimes grow on the carapace, as here, helping camouflage the turtle. The carapace is made up of hard shields called scutes.*

*When fully extended the **tail** is almost as long as the carapace. The tail can help the turtle right itself if it rolls over onto its back.*

*The **hind legs** do not fully retract into the shell.*

upper and the lower jaw are hooked at their front end to form a kind of beak. The upper jaw has an irregular outline. The skin forming the edges of the jaws is hard and very sharp. The jaws function rather like scissors (except that the two blades do not cross over one another): a big snapper can easily cut off a human finger with one bite.

The upper part of the shell of a turtle is called the carapace. It is covered with hard shields, or scutes. In alligator snapping turtles, each scute is conical and has a backward-facing

▼ *The common snapping turtle is a smaller animal than the alligator snapping turtle. Common snappers typically have a carapace 16 inches (40 cm) long, and the average weight of an adult is about 30 pounds (14 kg).*

Turtle plastrons

The plastron—the lower part of the shell—is covered by a row of large scutes on either side of the midline. A typical turtle has six of these scutes on each side. In some turtles, the scutes are hinged, so the plastron can move. Musk turtles and box turtles have a single hinge that runs between the pectoral and abdominal pairs of scutes. The underlying bones are also hinged. Mud turtles have two hinges, one between the pectoral and abdominal scutes and one between the abdominal and femoral scutes. Both the front and the back parts of the plastron can thus move. Some tortoises have a hinge between the abdominal and femoral scutes, but their shells are not as flexible as those of box turtles and mud turtles. Hinge-back tortoises from Africa also have a hinge, but only the back part of their carapace can move. Snapping turtles do not have hinges, and the plastron is very small. Most other turtles have a much shorter tail. No one really knows why snapping turtles have such a long tail, although a few years ago a clutch of captive snapping turtles hatched in which some had no tail. The tailless baby turtles could not right themselves so easily when they fell onto their back—a capacity essential for turtles. Nor could they climb steep slopes on land so well. However, this does not explain why not all turtles have a long tail.

▶ **Alligator snapping turtle**
The shell on the underside of a turtle is called the plastron. The cross-shaped plastron of an alligator snapping turtle is relatively small.

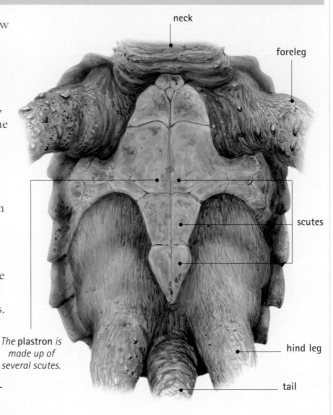

neck

foreleg

scutes

hind leg

tail

The plastron is made up of several scutes.

◀ *A pond turtle withdraws its head completely inside the shell, something that snapping turtles are unable to do. Pond turtles can achieve this because their head is small in relation to the size of their shell.*

point. These points form three ridges, or keels, that run along the length of the carapace. The matamata, a mud-dwelling turtle that lives in rivers in South America, has similar pointed scutes. In most turtles, including the common snapping turtle and the big-headed turtle, the scutes are flat and the carapace is smooth. This smooth profile makes the turtles streamlined and thus reduces drag during swimming.

The lower part of the shell of a turtle is called the plastron. The plastron of common snapping turtles and alligator snapping turtles is smaller than the plastron of most other species of turtles. The big-headed turtle, which may not be a close relative of the snapping turtles, has a large plastron.

Skeletal system

The boxlike shell that encases a turtle's body is unique among vertebrates. Turtles have flattened and enlarged ribs that support the top part of the shell, which is called the carapace. The carapace is made up of flat bony plates. The individual bones of the backbone are fused solidly to the ribs except at the front—where they form the neck—and the back, where they form the tail. There is a row of bones above the backbone. These are mostly called neurals. The bones around the edge of the carapace are called peripherals. Between them are the large ribs.

The bones that form the lower part of the shell—the plastron—have evolved from parts of the shoulder girdle and from ribs in the abdomen (such as occur in modern crocodiles). The carapace and the plastron are covered by scutes, or shields, which are derived from the

▼ Alligator snapping turtle
The plastron and the dorsal scutes have been removed in this ventral view. The costal bones do not reach the marginals; this arrangement reduces the weight of the upper shell, or carapace.

CLOSE-UP

Scleral ossicles

Many reptiles have a ring of flat bones called scleral ossicles surrounding the eye. The bones are particularly easy to see in snapping turtles. Alligator snapping turtles usually have about 12 scleral ossicles around each eye.

outer layers of the skin. The scutes lying above the backbone are called ventral scutes. There are fewer ventral scutes than neural bones, and each ventral scute is larger than a neural bone. The scutes at the edge are called marginals. In between the marginals and the ventrals are the large pleurals and the much smaller submarginals. Each turtle species has its own arrangement of scutes, although the variation among the 250 species is not great. The outlines of the scutes do not correspond to the outlines of the underlying bones.

COMPARE the neck bones of a snapping turtle with the backbone of a long, thin vertebrate such as a *GREEN ANACONDA* or a *GULPER EEL*. Snakes and eels have to be very flexible so they can bend when they are moving. The connections between the neck bones of turtles need to be equally flexible so the backbone can bend when the head is pulled backward into the shell for protection.

CONNECTIONS

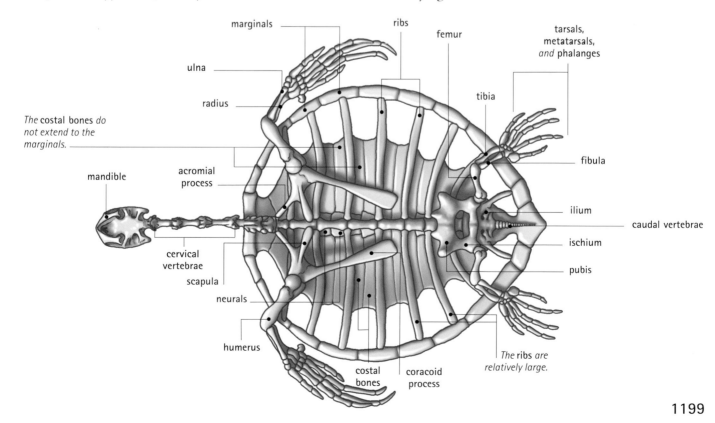

The costal bones do not extend to the marginals.

marginals · ribs · femur · tarsals, metatarsals, *and* phalanges · ulna · radius · tibia · fibula · mandible · acromial process · ilium · caudal vertebrae · ischium · cervical vertebrae · pubis · scapula · neurals · humerus · costal bones · coracoid process · The ribs are relatively large.

▲ *The skull of an alligator snapping turtle is relatively very large. The outline of the scleral ossicles—small, flat bones—are visible around the eye.*

All turtles and tortoises have a very solidly constructed skull. Alligator snapper skulls have two large spaces at the posterior end of their skull, which allow space for the massive jaw muscles. The muscles attach mainly to the squamosal and supraoccipital bones. The beak at the front end of the upper and lower jaw is also a prominent feature. Most other chelonians do not have a prominent beak.

The limbs of chelonians are short and stout. The bones of the fingers and toes are much longer in aquatic species than in terrestrial ones, especially in the front legs, because these bones support the flippers.

Evolution has greatly modified turtles' shoulder girdle. The main part of the shoulder girdle is formed by the two scapulae. In marine turtles, however, the coracoid bones are also well developed, because they are major attachments for the powerful muscles that the animals use for swimming. The shoulder girdle of chelonians has no scapulae: these bones have become incorporated into the structure of the plastron. The pelvic girdle is not so highly modified, and all of the bones found in other reptiles and mammals are present.

▼ BONES OF CARAPACE
Spiny softshell turtle

The carapace of snapping turtles and most other turtles is made up of bony plates, and the edge of the shell has a bony rim. In softshell turtles, there is no outer bony rim, but the shell is reinforced by cartilage and skin.

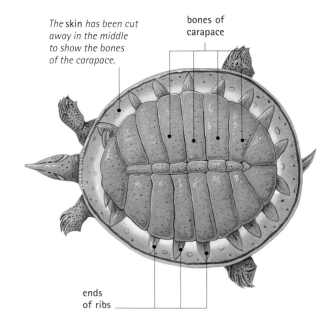

The skin has been cut away in the middle to show the bones of the carapace.

bones of carapace

ends of ribs

Muscular system

The muscular system of turtles reflects the way in which the body is highly modified by the shell. The trunk muscles are not large. They are not needed for support, which is provided by the shell, and the trunk region cannot move. The reduction in the size of the trunk muscles can be best appreciated by comparing a transverse cross section (across the body at right angles to the long axis) through the abdominal region with one just behind the neck. There are far more muscles in the neck. This is especially true in snapping turtles, which need powerful neck muscles to retract the large head. Most turtles can withdraw the head within the shell for protection. Snapping turtles cannot do this because their head is too big, but their neck can bend into an **S** shape to partially retract it. Because the neck movements of a snapping turtle are complex, the muscles are arranged in a complicated way.

In snapping turtles, the jaw muscles are huge. They enable the jaw to bite and grip with great force. Thus a snapping turtle can grab and break up large prey with its jaws. The muscles of the limbs are also large.

▼ SUPERFICIAL NECK MUSCLES
Chinese softshell turtle
Softshell turtles have a long neck that allows the nostrils to reach the water's surface when the body of the turtle is submerged. A long neck is also *useful for reaching out to grab prey when the turtle is lying in ambush. Like snapping turtles, softshell turtles can retract the neck into the shell for protection. Note the complex, intertwining pattern of the muscle groups.*

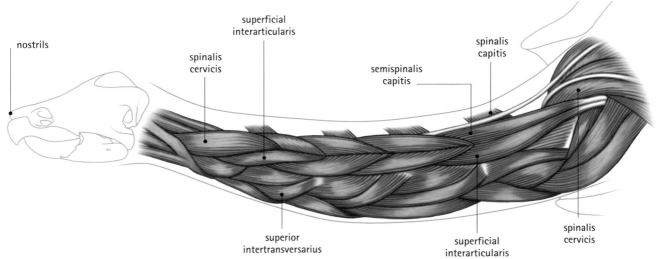

nostrils

superficial interarticularis

spinalis cervicis

semispinalis capitis

spinalis capitis

superior intertransversarius

superficial interarticularis

spinalis cervicis

Nervous system

Like all vertebrates, turtles have a nervous system that includes a central nervous system (CNS), consisting of the brain and spinal cord; and a peripheral nervous system (PNS), which consists of nerves connecting to the CNS. Except for a few very small details, a turtle's brain is very similar to the brains of lizards, snakes, and crocodilians. Turtles have fewer body segments than most other reptiles, and so they have fewer peripheral nerves.

Light sensors

In almost all vertebrate animals, the eyes are positioned on the side of the head. Light is detected on the eye's rear inner surface, which is called the retina. Most vertebrates can also sense light elsewhere on the body, but these light sensors, which are called extraretinal photoreceptors, are less well understood. One of the extraretinal photoreceptors in snapping turtles and other chelonians is a small knob on top of the brain called the pineal body. In turtles, as in many other vertebrates, the pineal body senses whether or not light is present and how strong its intensity is. Unlike the eyes, the pineal body does not form an image. The pineal body activates production of a

hormone called melatonin, which helps organize the rhythms of the body. The pineal body, however, is one of the least understood organs in reptiles and some other vertebrates.

IN FOCUS

Magnetic sense

People have wondered for years how marine turtles can spend 20 or more years wandering the oceans and then return to mate and reproduce at the beach where they hatched. It is now known that one of the methods turtles use to navigate is by using Earth's magnetic field. A compass works by lining up with the lines of polarity of the magnetic field: it shows the direction of the north (or south) magnetic pole. Turtles and birds use magnetic fields differently. They can sense the vertical angle of the magnetic field, which ranges from 90 degrees at the magnetic poles to 0 degrees at the equator. This ability was demonstrated by placing turtles in artificial magnetic fields and recording how they responded. Scientists do not know how turtles and birds can sense magnetic fields or how they use the information to find their way about. These animals do not, so far as it is known, have specialized sense organs to do this. It is likely that they can sense magnetic fields in parts of their brain, perhaps in some of the retinal cells of the eye. Some birds may also sense magnetic fields in some of the cells in their beak.

◄ Turtles have very good senses of sight and smell. Sensory information passes from the retina of each eye to the brain along the optic nerves, and information about odors passes from the olfactory epithelium, which is positioned behind the two nostrils.

Circulatory and respiratory systems

All turtles have a three-chamber heart that is very similar to the heart of a lizard or a snake. Oxygen-rich blood from the lungs passes via the pulmonary vein to the left atrium, and oxygen-depleted blood from the body passes through the sinus venosus to the right atrium. When the atria contract, blood is forced into the ventricle. When the ventricle contracts, blood is forced forward into the pulmonary arteries and the left and right aortic arches. The ventricle has two parts, but they are not completely separated from each other. A turtle is able to control the amount of blood that passes into the pulmonary arteries (which carry blood to the lungs) or the aortic arches (which carry blood to the body).

This ability is important for snapping turtles and other aquatic species because they cannot use their lungs for breathing when they are underwater. When this occurs, most of the

Lungs and buoyancy

Snapping turtles use their lungs to provide buoyancy when they are floating in water. The turtles even have the unusual ability to move air backward and forward within the lungs, so helping alter the angle of their body within the water.

blood from the ventricle is sent to the aortic arches; the lungs are effectively shut down because they are incapable of adding oxygen to the blood. Being able to control the flow of blood is also important for terrestrial species because they cannot breathe when the head is withdrawn into the shell.

COMPARE the heart of a reptile such as a turtle with the heart of a mammal such as an **ELEPHANT**, **PUMA**, or **WEASEL**. Turtles, lizards, and snakes have a three-chamber heart, whereas mammals have a heart with four chambers (two atria and two ventricles).

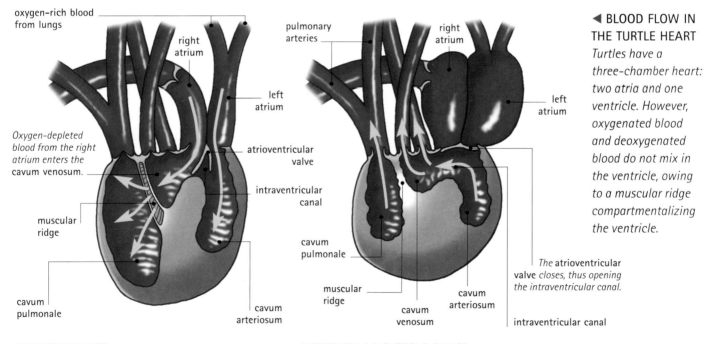

◄ BLOOD FLOW IN THE TURTLE HEART
Turtles have a three-chamber heart: two atria and one ventricle. However, oxygenated blood and deoxygenated blood do not mix in the ventricle, owing to a muscular ridge compartmentalizing the ventricle.

The atrioventricular *valve* closes, thus opening the intraventricular canal.

DIASTOLE PHASE
Oxygen-rich blood flows from the left atrium into the cavum arteriosum, and oxygen-poor blood from the right atrium enters the cavum venosum, then crosses the muscular ridge into the cavum pulmonale. The atrioventricular valve prevents the mixing of oxygen-rich and oxygen-poor blood.

VENTRICULAR SYSTOLE PHASE
When the ventricle contracts, oxygen-poor blood in the cavum pulmonale is pushed out through the pulmonary arteries. The atrioventricular valve closes, and oxygen-rich blood in the cavum arteriosum is forced into the cavum venosum. The muscular ridge now prevents the mixing of oxygen-rich and oxygen-poor blood.

IN FOCUS

How turtles survive

Mammals are able to shiver to help generate heat when they are cold. Turtles (and other reptiles) cannot do this. Many turtles in colder climates, however, must be able to survive cold winters. This sometimes involves withstanding temperatures that are below freezing. Some turtles in the northern parts of North America can do this by converting a complex carbohydrate called glycogen in the liver to a simple carbohydrate called glucose. The glucose is then carried by the blood to tissues, where it helps prevent the cells from freezing. Box turtles and painted terrapins can survive for many days at temperatures of nearly 25°F (−4°C), and they can withstand more than half of their total body water turning to ice.

▼ This alligator snapping turtle stands motionless in water, awaiting prey. The pink lure on its tongue is visible. Aquatic turtles can obtain oxygen from water through the lining of their mouth and throat.

Organs for gas exchange

All turtles and tortoises have a pair of large lungs, which lie at the top of the body cavity beneath the shell. Aquatic forms cannot breathe using lungs when they are underwater. They have extra mechanisms that enable them to remain submerged for long periods. Their blood can carry a lot of oxygen; their tissues use the oxygen only very slowly; and they can switch to a metabolism that does not require oxygen. This is called anaerobic respiration, and it can be sustained only for relatively short periods.

Turtles can add to the oxygen uptake they receive from the lungs with supplementary respiratory mechanisms elsewhere in the body. There are three places where supplementary gas exchange occurs, but the extent to which they are used varies from species to species. Gas exchange may occur through the skin and across the lining of the mouth and throat: some species of turtles pump water there by making swallowing movements when they are underwater. Additionally, many aquatic turtles have a pair of thin-walled anal sacs, one on either side of the cloaca, through which gas exchange occurs; like all respiratory organs, these are well supplied with blood vessels.

Digestive and excretory systems

Snapping turtles eat mostly meat: fish, amphibians, and mollusks all feature in their diet, as does some plant matter. Young snapping turtles pursue their prey more energetically than adults. Unlike most other chelonians, which have a simple, short tongue that cannot be protruded from the mouth, alligator snapping turtles have a tongue with a long, thin, red projection on the top. The turtle can move this projection so it wriggles like a worm. The moving projection is used as a lure to entice prey within range of the powerful jaws. Scientists have filmed snapping turtles feeding. If the film is slowed down it shows that when the turtle eats a small fish, the entire process of snapping it into the jaws takes only 78 milliseconds—less than $\frac{1}{10}$ of a second. It was once thought that prey were sucked into the open mouth, but the same studies showed that the snapper projects its head forward with the mouth open, so engulfing the prey.

All turtles have a straightforward digestive system with a simple one-chamber stomach. The main variation between species is in the size of the large intestine. This is larger and longer in species such as terrestrial tortoises that eat mostly plants, because the large intestine must have space to house a large population of bacteria, which helps digest the thick cellulose in the walls of plant cells. Most freshwater turtles, including snapping turtles, are omnivorous; that is, they eat both plants and animals. Thus their large intestine does not need to be so long.

As in all reptiles, the last part of the alimentary canal is a specialized region where the digestive, excretory, and reproductive systems meet. This is called the cloaca. The rectum, the ureters carrying urine from the kidneys, and the ducts carrying spermatozoa or eggs from the gonads all meet at the cloaca. Most freshwater turtles also have two anal sacs, one on each side of the cloaca. These sacs are supplementary respiratory organs.

▼ A fish sees an alligator snapping turtle's lure and approaches, perhaps believing that the lure is a worm. If the fish comes close enough, the turtle will move its head forward and slam shut its jaws.

Reproductive system

COMPARE the single penis of a male snapping turtle with the two hemepenises of a male *GREEN ANACONDA* or a male *JACKSON'S CHAMELEON*. Although male snakes and lizards have two penises, only one is used during mating.

The fertilization of a female snapping turtle's eggs is internal. When a pair of turtles is ready to mate, the male uses its claws to grip the female's carapace as he climbs onto her. The male bends his tail until the opening of his cloaca touches hers, then he inserts his penis into the female's cloaca (reproductive chamber) and ejects sperm. Male turtles have only one penis, situated on the dorsal (top) part of the cloaca. All male snakes and lizards—the majority of modern reptiles—have two penises, one on either side of the cloaca.

The sperm may fertilize the female's eggs, which are produced in her ovaries. The ovaries of alligator snappers are very colorful in summer before the eggs are laid, although the reason for this is unknown. The ovarian tissue is dark blue or purple, and the developing fertilized eggs are yellow. A white structure in each ovary, called the corpus luteum, releases the reproductive hormone progestorone.

Reproductive pattern

Common snapping turtles are able to reproduce when their carapace is around 8 inches (20 cm) long. Common snappers mate between May and November, whereas their close relatives, alligator snapping turtles, have a much more restricted mating period of March and April.

Snapping turtles do not have elaborate courtship behavior, but the males of many other freshwater turtles do. In painted turtles such as pond sliders, the male faces the female and strokes her head with the claws on his front feet. This often happens on the surface of the water, but the pair then sink to the bottom to mate. Mating may take more than an hour and is sometimes interrupted by the necessity for one partner or the other to come to the surface for a gulp of air.

Adult female common snapping turtles lay a clutch of 20 to 40 eggs, and alligator snappers lay up to 50 eggs in a nest that is excavated on land but close to freshwater. The almost spherical eggs hatch after about three months. The eggshell is softer and more pliable than that of a bird's egg. Turtle eggs can be dropped from a height of more than 3 feet (1 m) onto concrete without breaking.

Alligator snapping turtle hatchlings emerge from the nest in early fall, 100 to 140 days after the eggs are laid. The hatchlings have to fend for themselves, and many fall victim to predators, including waterbirds.

Sex determination

In many turtles, the sex of hatchlings is determined not genetically but by the temperature of the surrounding soil or sand

► UROGENITAL SYSTEM VIEWED FROM ABOVE
Male alligator snapping turtle
Male turtles have two sperm-producing testes and one penis. Females have two egg-producing ovaries. The rectum opens into the cloaca, as do the bladders.

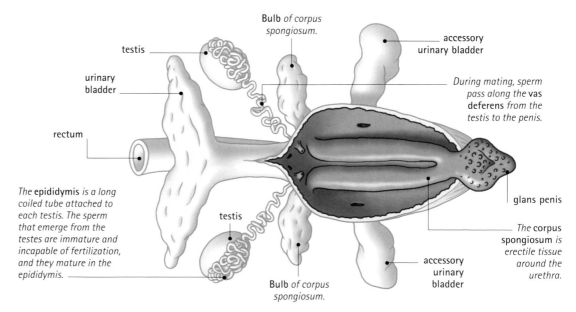

testis

urinary bladder

rectum

The epididymis is a long coiled tube attached to each testis. The sperm that emerge from the testes are immature and incapable of fertilization, and they mature in the epididymis.

testis

Bulb *of corpus spongiosum.*

Bulb *of corpus spongiosum.*

accessory urinary bladder

During mating, sperm pass along the vas deferens *from the testis to the penis.*

glans penis

The corpus spongiosum *is erectile tissue around the urethra.*

accessory urinary bladder

while the eggs are incubating. The most common pattern is for clutches that have been exposed to cool temperatures to produce all males, whereas those that have been exposed to hot temperatures produce all females, and those exposed to intermediate temperatures produce a mixture of sexes. In alligator snapping turtles, however, the system is more complicated. Low and high incubation

▲ *A common snapping turtle that has recently broken out of its egg. When hatchlings first emerge in May or June, they are just 1.5 inches (3.8 cm) long.*

temperatures produce all males, whereas intermediate temperatures produce mostly females. There is no temperature, however, at which only females are produced. The critical time for determining the sex of offspring is just before the halfway point in the development of the embryo. This is when the cells that will develop into gonads (testes or ovaries) have just begun to appear.

There are other reptiles in which the sex of offspring is determined by temperature. For example, in many lizards high temperatures produce only male lizards; this situation is the opposite of what occurs in many turtles. Most snakes, however, have a genetic sex-determining system. Individuals with two identical sex chromosomes become males, whereas those with two different sex chromosomes become females. This setup is the reverse of the system in mammals.

ROGER AVERY

IN FOCUS

Several fathers

Female turtles of many species, including common and alligator snapping turtles, can store living sperm in their oviducts for many months. In theory, therefore, the turtles could mate many times, so that some of the eggs in a clutch have been fertilized by sperm from one father and others by sperm from another. Studies of common snapping turtles in Canada have found that this indeed happens. DNA fingerprinting was used to study the genetic composition of all the offspring from a single clutch in three females. In two of the three clutches the genetic variation in the offspring was too great to be the result of mating with only one father; the offspring had resulted from matings with two, possibly three, fathers. In some species of turtles, the females can store sperm for several years. These species include the diamond-backed terrapin and the eastern box turtle.

FURTHER READING AND RESEARCH

Halliday, T., and K. Adler. 2002. *The New Encyclopedia of Reptiles and Amphibians*. Firefly: Toronto.

Harris, T. (ed.). 2003. *Reptiles and Amphibians*. Marshall Cavendish: Tarrytown, NY.

Orenstein, R., G. Zug, and J. Mortimer. 2001. *Turtles, Tortoises, and Terrapins*. Firefly: Toronto.

Squirrel

ORDER: Rodentia FAMILY: Sciuridae
SUBFAMILY: Sciurinae GENERA: 50

Squirrels are among the most widespread of mammals and live in a diverse range of habitats. From tropical rain forests and semiarid deserts to coniferous forests and even temperate city gardens, squirrels have been able to exploit many ecological niches. Many species of squirrels have an arboreal (tree-living) lifestyle, but there are also ground-living species that live in burrows.

Anatomy and taxonomy
Scientists group all organisms into taxonomic groups based largely on anatomical features. Scientists place squirrels in the order Rodentia—the rodents—which are among the most numerous and successful of all animal groups. Squirrels belong to a group of rodents that also includes chipmunks, marmots, and prairie dogs.

● **Animals** All animals are multicellular. They get the energy and materials they need to survive by consuming other organisms. Unlike plants, fungi, and the members of other kingdoms, animals are able to move around for at least one phase of their life.

● **Chordates** At some time in its life cycle, a chordate has a stiff dorsal (back) supporting rod called a notochord. Most, although not all, chordates are vertebrates. The notochord of vertebrates becomes part of the spine, or backbone. The spine is made up of units called vertebrae, which are generally made of bone.

● **Mammals** One of the eight classes of vertebrates, mammals are warm-blooded animals with four limbs and, in most cases, a tail. They have body hair, which generally covers nearly all of the body surface, making a thick fur. All mammals nourish their newborns with milk secreted from

▼ *There are three main types of squirrels: tree squirrels, ground squirrels, and flying squirrels. This family tree shows that the gray and red squirrels are members of the family Sciuridae, which are mammals in the order Rodentia.*

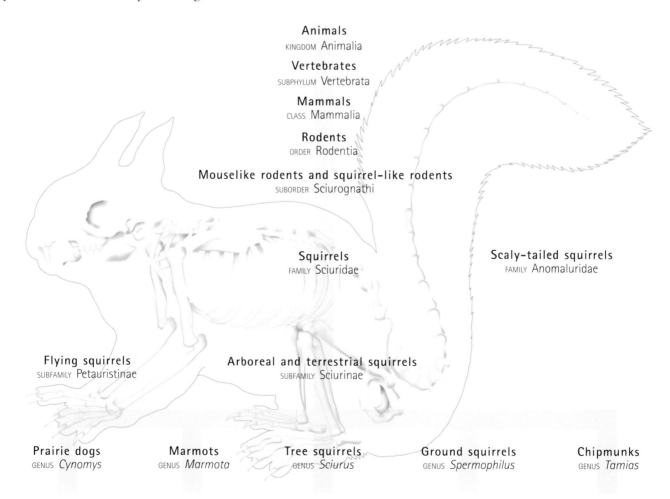

Animals
KINGDOM Animalia

Vertebrates
SUBPHYLUM Vertebrata

Mammals
CLASS Mammalia

Rodents
ORDER Rodentia

Mouselike rodents and squirrel-like rodents
SUBORDER Sciurognathi

Squirrels
FAMILY Sciuridae

Scaly-tailed squirrels
FAMILY Anomaluridae

Flying squirrels
SUBFAMILY Petauristinae

Arboreal and terrestrial squirrels
SUBFAMILY Sciurinae

Prairie dogs
GENUS Cynomys

Marmots
GENUS Marmota

Tree squirrels
GENUS Sciurus

Ground squirrels
GENUS Spermophilus

Chipmunks
GENUS Tamias

mammary glands on the female's underside or front. Three species of mammals lay eggs, and some others suckle their young in pouches on the underside of the body. Most mammals, including squirrels, nourish their unborn young through a temporary organ called the placenta. This organ allows the young to develop while still inside the mother. Mammalian red blood cells lack a nucleus.

- **Rodents** This group is a very large order of placental mammals. Compared with other mammals, most of the 2,000 species of rodents are small. They are equipped with large chisel-like incisors, which make these mammals expert gnawers. This ability has allowed rodents to exploit a huge variety of foods and live in a wide array of habitats.

- **Sciurognathi** This suborder of rodents includes the squirrels, prairie dogs, beavers, and mouselike rodents, such as rats. Sciurognaths are distinguished from other rodents, which include guinea pigs (or cavies), chinchillas, capybaras, and porcupines. Members of this second group—the cavylike rodents—have a larger head and a more robust body than sciurognaths. The sciurognaths are thought to resemble more the primitive rodent form from which all modern rodents evolved.

- **Sciuridae** This family includes the arboreal and terrestrial squirrels but does not include the scaly-tailed squirrels, which belong to the family Anomaluridae. The family Sciuridae contains more than 270 species and is divided into two subfamilies: the Sciurinae and the Pteromyinae. The Pteromyinae includes the flying squirrels.

- **Sciurinae** This subfamily of rodents contains the tree squirrels, ground squirrels, chipmunks, and marmots. Their

▲ The coloration of the red squirrel varies from bright reddish brown to dark grayish brown or almost black.

range covers most continents except Australia, southern South America, and some desert regions.

- *Sciurus* The American tree squirrels and the Eurasian tree squirrels that are found north of the tropics belong to the genus *Sciurus*. This geographically widespread genus has a total of 28 species, including the Eurasian red squirrel, Arizona gray squirrel, Japanese squirrel, and yellow-throated squirrel.

FEATURED SYSTEMS

EXTERNAL ANATOMY Squirrels are medium-size rodents with a small, rounded head and a long, bushy tail. Squirrels have dexterous front paws that they use to manipulate food and, in the case of tree squirrels, use for climbing. *See pages 1210–1213.*

SKELETAL SYSTEM The skeleton of a squirrel is relatively light, allowing the squirrel to climb and leap easily from tree to tree. Squirrels walk on the soles of the feet and have long toes terminating in claws. *See pages 1214–1215.*

MUSCULAR SYSTEM The tail muscles enable a squirrel to move its tail in any direction. Flying squirrels have a muscular gliding membrane called a patagium. *See pages 1216–1217.*

NERVOUS SYSTEM Squirrels have an excellent sense of smell and use scent to communicate with other squirrels.

They also have very good eyesight, which tree squirrels use to navigate accurately from branch to branch. *See page 1218.*

CIRCULATORY AND RESPIRATORY SYSTEMS Like all other mammals, squirrels have a four-chamber heart and a closed circulatory system consisting of arteries and veins. Respiratory gases are transferred in and out of the body across the surfaces of a pair of lungs. *See page 1219.*

DIGESTIVE AND EXCRETORY SYSTEMS The digestive system is suited to a diet made up mostly of plant matter and has a large cecum containing bacteria that break down cellulose. *See pages 1220–1221.*

REPRODUCTIVE SYSTEM Squirrels have one or two litters each year, depending on the species. There are usually one to six pups in each litter. *See pages 1222–1223.*

External anatomy

COMPARE the hind feet of a squirrel with those of a **HARE**, which has hind feet that are suited to leaping rather than climbing and cannot be rotated at the ankle.

COMPARE the tail of a squirrel with that of a **JACKSON'S CHAMELEON**. The reptile's tail is fully prehensile (able to grip), whereas that of the squirrel is used only for balance.

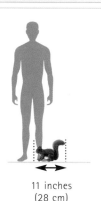

11 inches
(28 cm)

▶ Eurasian
red squirrel
This species of squirrel is easily recognized by its reddish brown fur, large, dark eyes, prominent ear tufts, rounded head, bushy tail, and long facial whiskers. Adults of this species weigh 9 to 11.3 ounces (250–320 g).

Squirrels are expert climbers and diggers. They are also among the most successful mammalian groups in terms of the number of species: there are more than 270 species worldwide. They have evolved to suit life in a wide variety of habitats. Many species live in forests, from the tropics to the cool temperate zone. Other species inhabit deserts, plains, or tundra. A few species, particularly the eastern gray squirrel, have even adapted to life in city parks and backyards. The smallest species is the mouse-size African pygmy squirrel, with a head and body length of just 2.6 to 3.9 inches (6.6–10 cm). The largest species is the heavyweight alpine marmot, which is 20.8 to 28.7 inches (53–73 cm) long. The biggest tree squirrel is the black giant squirrel of southeastern Asia, which has a head and body length of up to 18 inches (46 cm) and a tail almost as long again.

*Prominent tufts of fur give the **ears** a pointed appearance.*

*The **eyes** are large and dark, set high and slightly to the sides of the head. They give a wide field of view without impairing the vision needed to see in stereo.*

*The **whiskers** are long and sensitive, helping squirrels "feel" their way as they move rapidly along branches.*

*The **nose** provides an excellent sense of smell, which helps the squirrel detect even food that is buried underground.*

*The **claws** are robust and sharp, anchoring the squirrel as it scales or descends steep angles on tree trunks and branches.*

*The **forelimbs** are stout and used for grasping food and climbing.*

Footpads

All squirrels have hard pads of skin on the soles of their feet that provide a better grip for holding food and extra traction when the animal is in motion. The long footpads of the long-clawed ground squirrel are covered by fur, which protects the feet from burning when the animal is running over hot desert sand. The outer edges of the hind feet also have fringes of stiff hair that help push the sand away in burrowing.

Body shapes

Typical tree squirrels, such as the Eurasian red squirrel, have a mainly arboreal lifestyle, spending much of their time high in the trees. Tree squirrels have a long, supple, cylindrical body, a rounded head with prominent ears, and a large bushy tail. Squirrels have short forelegs and longer hind legs. They can descend trees headfirst and rotate their double-jointed hind feet backward while their claws dig into the bark for support.

Ground-dwelling squirrels, such as the arctic ground squirrel, are heavier-bodied with shorter legs, ears, and hair and a less bushy tail than tree squirrels. These features help squirrels pass in and out of their burrows with ease. Their strong forelimbs have large claws for scratching and digging in the soil. Both tree and ground

*The **tail** is a versatile appendage providing balance, warmth, or shade and is essential in communicating with other squirrels.*

*The **fur** is soft, is richly colored, and grows much thicker during winter to provide better insulation from cold weather.*

*Flexible **ankle joints** enable the hind feet to rotate backward, allowing the squirrel to dangle from its hind feet or descend a tree trunk headfirst.*

▲ SIZE COMPARISON
The black giant squirrel (right) of southeastern Asia is much larger than the Eurasian red squirrel and can reach 18 inches (46 cm) from nose to rump, with a tail as long as its body or even longer.

COMPARATIVE ANATOMY

Head shapes

Most squirrels typically have a rounded head with large eyes, a short snout, and prominent front teeth, which are used for gnawing food. Some ground squirrels have cheek pouches for storing food. Nocturnal flying squirrels have larger eyes and ears; well-developed senses help these squirrels live in darkness. Unusually, the aptly named shrew-faced ground squirrel from southeastern Asia has a long, pointed muzzle similar to that of the tree shrews, but its short, bushy tail helps distinguish it from these rodents.

▶ EAR SHAPES

Good hearing is vital for flying squirrels because they are active at night and thus less able to rely on sight. The ears of the gray squirrel lack the tufts of the red squirrel, but gray squirrels are equally sensitive to sounds of danger and the calls of other squirrels.

Giant flying squirrel   Gray squirrel

▲ Ground squirrel

Ground squirrels in the genus Spermophilus *inhabit open country in North America, eastern Europe, and central Asia. The name* Spermophilus *means "seed-lover," reflecting the animal's preferred diet.*

squirrels have a small thumb and four toes on the forelegs and five toes on the hind legs, with the exception of the woodchuck of North America, which has just four hind toes.

Ears and eyes

Chipmunks, tree squirrels, and flying squirrels all have large ears that are highly sensitive to sound. The Eurasian red squirrel also has distinctive tufts of fur that give its ears a pointed appearance, a feature shared by several other species, including its North American relative, Abert's squirrel; and the groove-toothed squirrel, which lives only in Borneo. All squirrels have large eyes circled by a ring of pale fur. The eyes are positioned high on the sides of the head. They give a wide field of view and binocular vision, which enables squirrels to judge distances with great accuracy.

Multipurpose tail and sensitive hair

A squirrel's bushy tail has several important functions. It acts as a rudder when the squirrel leaps, provides essential balance during climbing

and running, and serves as a signaling device to other squirrels, particularly when danger threatens. A squirrel's tail is 80 to 90 percent the length of its head and body and can thus provide vital warmth when curled around the body during sleep. Similarly, when draped across the back it shades the animal from the sun during hot weather.

Like other squirrels, the Eurasian red squirrel has highly touch-sensitive hairs called vibrissae on its head, feet, and outer legs, and at the base of its tail. The extra sensitivity these hairs give helps the squirrel navigate quickly and efficiently as it moves through the trees. The Eurasian red squirrel molts its body fur twice each year, in spring and fall; but the tail and ear tufts are molted only once, during the summer. Its fur is usually rich chestnut on most of the body with pale cream fur on the underside, but there is much regional variation, and color may vary from light brown to almost black. The fur of a typical Eurasian red squirrel becomes thicker and darker during the winter, and in Russia this species is still hunted for its attractive pelt during the cold months.

patagium (gliding membrane)

▲ **Giant flying squirrel**
The gliding membranes on the flanks of the giant flying squirrel are tucked away when the animal climbs along branches.

▲ *Although a flying squirrel, such as this small Japanese flying squirrel, does not have the advantage of powered flight, it can glide through the forest air with grace and precision. The fur-covered gliding membrane is called a patagium.*

Leapers and gliders

Although bats are the only mammals that can truly fly, flying squirrels have evolved over millions of years the ability to glide from tree to tree using a parachute-like membrane of furry skin stretched between their long forelegs and hind legs. This gliding is a very efficient form of locomotion: giant flying squirrels in Asia can travel as far as 1,500 feet (460 m) in a single glide. Flying squirrels have large, dark eyes, dense, soft fur, and a long, flattened tail, which they use to steer as they glide. They are highly arboreal: they spend most of their time in trees and rarely descend to the ground. One of the best-known species, the southern flying squirrel, is found across much of the eastern United States. Like most flying squirrels, however, it is strictly nocturnal and therefore difficult to observe. Biologists believe the gliding membranes make these squirrels more vulnerable to predators during daylight, so nocturnal habits enable them to avoid danger.

The scaly-tailed flying squirrels of Africa are not considered true squirrels. The anatomy of their head differs significantly from that of other flying squirrels, and their gliding membranes are attached at the elbow rather than at the wrist. They also have a shorter, tufted tail. The underside of the tail has twin rows of raised scales that are used for gaining extra purchase during climbing, and the tail serves as a rapidly deployed anchor when the squirrel lands.

◄ *Sharp hind claws, flexible ankle joints, and a balancing tail enable tree squirrels, such as this Eurasian red squirrel, to perch on branches with poise and climb with extraordinary agility.*

1213

Skeletal system

COMPARE the plantigrade feet of a squirrel with the digitigrade feet of a *PUMA*. Locomotion in plantigrade animals involves walking or bounding with the soles of the feet touching the ground, whereas digitigrade animals use just the tips of their toes for walking or running.

CONNECTIONS

In common with all other mammalian skeletons, a squirrel's skeleton can be divided into two parts: the axial skeleton, which is the skull and spine from neck to tail; and the appendicular skeleton, which consists of the limbs, pectoral girdle, and pelvic girdle. A tree squirrel's skeleton typically has lightweight bones with flexibility that allows the squirrel to leap and climb easily. Caudal vertebrae form a long, slender, and very flexible tailbone, giving the tail greater mobility for balancing, signaling, and providing protection from the elements.

The hind limb bones are longer, denser, and heavier than those of the forelimbs, giving the squirrel extra strength and support as it moves. The toes are long. Squirrels are plantigrade mammals: they bound with the soles of their feet touching the ground. The thumb bones on the forefeet are much smaller, but all digits are equipped with strong, curved claws. A squirrel's wrist bones are strong but flexible, enabling the forelimbs to be used constantly during foraging and eating. Ground squirrels also use their forelimbs to assist in digging; tree squirrels use them for climbing; and flying squirrels use them for adjusting the shape and size of the gliding membrane during glides.

A rounded skull

A Eurasian red squirrel's skull is rounded and has deep eye sockets, which accommodate large eyes. Like other squirrels, the red squirrel has a

▼ **Eurasian red squirrel**
The red squirrel's skeleton is a light, flexible frame that enables the animal to move easily in the trees. The hind limbs are longer than the forelimbs and propel the animal during running or climbing. The long tail helps balance the squirrel.

caudal vertebrae

cervical vertebrae

The scapula *is part of the pectoral girdle.*

thoracic vertebrae

sacral vertebrae

lumbar vertebrae

eye orbit

cranium

zygomatic arch

dentary

sternum

humerus

ulna

radius

phalanges

metacarpals

carpals

ribs

tibia

fibula

femur

pelvic girdle

tarsals, metatarsals, and phalanges

lower jaw structure that is relatively primitive but is also strong and mobile. The lower jaw juts forward when the squirrel gnaws.

Gnawing teeth

Squirrels' teeth are typical of rodents in their adaptation for gnawing. Squirrels have a single pair of rootless, chisel-like incisors in each jaw, and one or two premolars and three molars on each side of each jaw. Squirrels do not have canines, teeth that are highly developed in carnivores for tearing meat.

There is a large gap between the incisors and premolars, which is called the diastema. Squirrels can suck their lips into this space to prevent nutshells and other inedible debris from being swallowed as the squirrel gnaws. Unlike the incisors, the cheek teeth have roots, and their crowns are low and covered by a rough ridged surface that helps the squirrel grind up hard food such as nuts.

EVOLUTION

Determining relationships

Scientific examination of fossil and modern skeletons has shown that squirrels have changed little in millions of years. The pattern of evolution has made the determination of their family tree difficult. The earliest known squirrel, called *Protosciurus*, appeared in the late Oligocene epoch about 34 million years ago, and the fossil remains of its skeleton suggest that it was a tree climber.

Some scientists believe that flying squirrels evolved separately from other squirrels, and their ancestors were the paramyid rodents of the early Eocene period, about 54 million years ago. These large, primitive rodents had squirrel-like features, such as a long tail and clawed feet. This theory is based on studies of fossilized teeth, but increasingly other biologists argue that the evidence is incomplete and look instead to modern genetic studies to trace the origins and relationships within the squirrel family tree. Consequently, many experts now believe that flying squirrels are a subfamily of the family Sciuridae, having evolved from the same common ancestor as modern tree squirrels.

CLOSE-UP

Incisors

Like the incisors of all other rodents, a squirrel's incisors have no roots and a high crown (such high-crowned teeth are termed hypsodont). They grow constantly throughout the animal's life and must therefore be worn down by constant use to prevent them from growing too long. The rear surface of the incisors has no enamel coating, so as the teeth grind against each other when the squirrel gnaws, the softer dentine layer on the rear surface is eroded. The enamel is then exposed as the chisel-like cutting edge of the tooth, which is highly effective at cutting open the hard outer shells of nuts and seeds.

▶ Typical of squirrels and rodents in general, this arctic ground squirrel has large chisel-like incisors. The beveled edge of the teeth results from the erosion of dentine on the incisors' rear surface.

Muscular system

COMPARE the structure of the patagium of a flying squirrel with that of a **FRUIT BAT**. The flying squirrel's patagium is a fur-covered muscular membrane used for gliding, whereas that of the fruit bat is thinner, with fine hairs, and is used for true flight.

A squirrel's musculature can be divided into three types: the cardiac muscles of the heart; the smooth muscles that enclose the blood vessels and the digestive and excretory systems; and the skeletal, or striated, muscles that are attached to the bones and enable the squirrel to move. Skeletal muscles are arranged in opposite, or antagonistic, pairs. When one muscle (the extensor) contracts, it causes a movement in one direction. The contraction of the opposing muscle (the flexor) results in movement in the opposite direction. The action of skeletal muscles is essential for all the squirrel's movements, from climbing, leaping, and gnawing to digging and blinking.

Jaw muscles

Squirrels have a distinctive arrangement of jaw muscles that enables them to gnaw their food. The principle jaw muscle is the masseter, which has several branches and is responsible for moving a squirrel's lower jaw as it gnaws and chews. The lateral branch stretches in front of the eye to the snout and is anchored by way of a broad plate on the skull. It is responsible for directing the movement of the incisor teeth by pushing the lower jaw forward when the squirrel gnaws. The superficial branch is much shorter and is used only for closing the jaws. The lateral branch of the masseter is attached to the skull by way of a

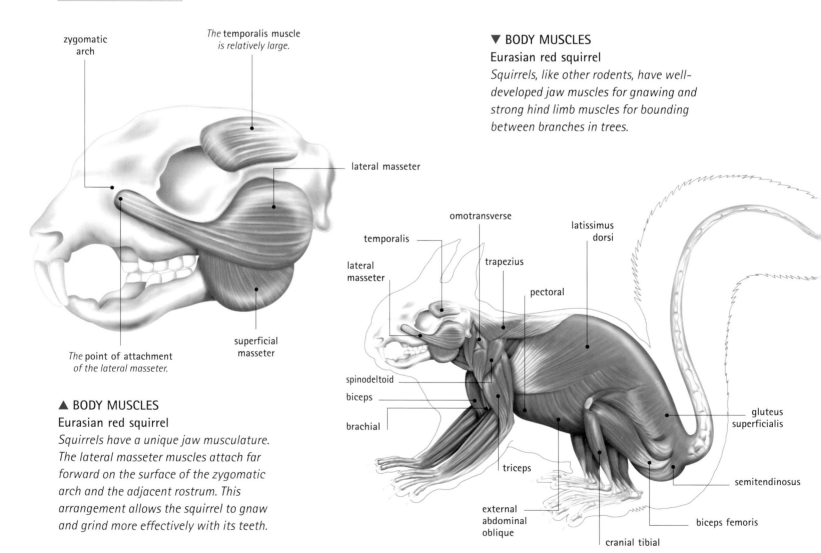

zygomatic arch

The temporalis muscle *is relatively large.*

lateral masseter

superficial masseter

The point of attachment *of the lateral masseter.*

▲ BODY MUSCLES
Eurasian red squirrel
Squirrels have a unique jaw musculature. The lateral masseter muscles attach far forward on the surface of the zygomatic arch and the adjacent rostrum. This arrangement allows the squirrel to gnaw and grind more effectively with its teeth.

▼ BODY MUSCLES
Eurasian red squirrel
Squirrels, like other rodents, have well-developed jaw muscles for gnawing and strong hind limb muscles for bounding between branches in trees.

omotransverse

temporalis

lateral masseter

trapezius

latissimus dorsi

pectoral

spinodeltoid

biceps

brachial

triceps

external abdominal oblique

cranial tibial

gluteus superficialis

semitendinosus

biceps femoris

knob of bone on the rostrum (front of the skull) called the masseteric tubercle.

As the squirrel chews and gnaws, the masseter is assisted by two additional sets of jaw muscles: the pterygoid and temporal muscles. The pterygoid muscles generate horizontal movement, and the temporal muscles create vertical movement. The pterygoid's flexibility helps the squirrel grind its food effectively. The arrangement and structure of jaw muscles are remarkably similar in most species of squirrels, but scientists have discovered that pygmy tree squirrels have unusually small temporal muscles and a superficial masseter muscle positioned in such a way that it allows the jaw to retract.

Muscles for gliding

Flying squirrels have a fur-covered muscular gliding membrane called a patagium, which acts like a parachute. The patagium looks like a flap of loose skin running continuously from the front legs to the hind legs—in some species it is also attached to the neck and tail. The patagium is made up of thin sheets of muscle that can be tensed or relaxed at will. At the front it is supported by a thin rod of cartilage on each side, attached to the wrists. When

airborne, the squirrel steers with great accuracy by varying tension in the patagial muscles and changing the position of its limbs and tail. A flying squirrel is even able to maneuver abruptly at a right angle to a branch at the last moment before landing.

The muscular tail

Squirrels have superb muscular control over their tail and are able to maneuver it in any direction for balance, communication, and protection. When threatened, a squirrel will rapidly flex its tail muscles to flick the tail from side to side and erect the long brush hairs, perhaps to distract the enemy and warn other squirrels of danger. If the tail is held by a predator, it can break off, allowing the squirrel to escape. Any exposed muscle and vertebrae soon dry up and are then shed. The animal can usually survive despite the loss of its tail.

A flying squirrel uses its tail as an effective brake just before landing, by suddenly flexing the muscles: the tail curves upward at the end of a glide, and the squirrel's body immediately follows suit. The velocity of the glide is then considerably reduced just before contact with a tree trunk. The place from which the squirrel launches its glide is always higher than the place at which it lands. The tail is also used for balance as the squirrel then climbs the tree.

◀ *Just before contact with a tree trunk, the southern flying squirrel raises its tail by flexing muscles at the base. In this position, the tail acts as a brake, lessening the impact on landing.*

Nervous system

Within the nervous system, there are two branches working together: the central nervous system (CNS; brain and spinal cord) and the peripheral nervous system (PNS; nerve fibers branching from the CNS). The CNS controls the voluntary actions of the body, such as skeletal-muscle movement. The PNS controls the involuntary movements and regulates heartbeat, the movement of smooth muscles in the digestive system, and the glandular release of hormones. The last is the trigger for the rapid responses of "fight or flight," flight being particularly important when the squirrel needs to escape from a predator.

Dichromatic eyesight

The visual sensitivity of most mammals is concentrated in a small area of the retina called the fovea. However, squirrels have the advantage of equal sensitivity across their entire retina, which gives them excellent eyesight for seeing food and predators, such as hawks and martens. Densely packed cones in the retina also provide very good dichromatic color perception: squirrels can distinguish most colors except red and green.

The endocrine system

The nervous and endocrine systems work together to control and regulate a squirrel's body activities and senses. The endocrine system affects growth, development, tissue function, and metabolic and reproductive processes. It is a network of glands that produce hormones, which are secreted directly into the squirrel's bloodstream and reach all regions of its body. Exocrine glands play an important role, as they produce scent secretions. Squirrels have an excellent sense of smell and use scent as a means of communication. The exocrine glands are located on the feet and are used to mark territory or leave other chemical messages, such as when a female comes into estrus and is ready to mate. Some species also scent-mark branches by wiping them with secretions from inside the mouth.

▶ **Eurasian red squirrel**
Squirrels have four sets of sensitive whiskers, or vibrissae, on the head. They are located above and below their eyes, in front of the throat, and alongside the nose and act as touch receptors that help relay information about the squirrel's immediate surroundings to the brain. There are also touch-sensitive whiskers on the wrist, at the base of the tail, and around the feet.

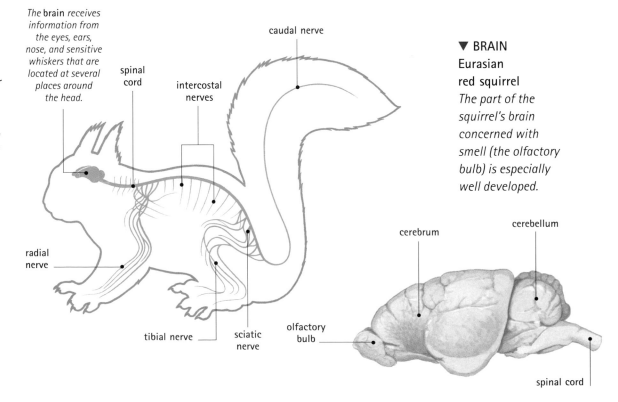

The brain receives information from the eyes, ears, nose, and sensitive whiskers that are located at several places around the head.

spinal cord

intercostal nerves

caudal nerve

radial nerve

tibial nerve

sciatic nerve

olfactory bulb

cerebrum

cerebellum

spinal cord

▼ **BRAIN Eurasian red squirrel**
The part of the squirrel's brain concerned with smell (the olfactory bulb) is especially well developed.

Circulatory and respiratory systems

The squirrel's circulatory system is essential for survival. It sends oxygen from the lungs to cells throughout the body and sends soluble nutrients from the small intestine to the cells. The circulatory system also transports important antibodies and hormones, pumps water from the cells to the kidneys, and takes carbon dioxide to the lungs, where it is released from the body as the squirrel exhales.

The heart and arteries

The driving force of the circulatory system is the heart. The squirrel's heart is a typically mammalian four-chamber structure and is made of cardiac muscle. Blood is pumped from the heart to the lungs, where oxygen is collected, and carbon dioxide is expelled. The blood then returns to the heart to be pumped around a larger circuit taking in the organs and other parts of the body. It travels in large arteries, which are long, robust tubes with strengthened walls to cope with the pressure of blood being pumped around the body. The arteries are connected to a network of ultrafine tubes called capillaries, which have thin walls and are in direct contact with the body's cells. Capillaries facilitate the exchange of both beneficial materials and waste. The blood is then drained back through veins to the heart, where the cycle begins again.

Respiration

Squirrels share a very similar respiratory structure with other rodents, such as rats, mice, and jerboas. A squirrel's respiratory system consists of lungs with alveoli and air passages—bronchioles, bronchi, the trachea (windpipe), and the nasal passages. The lungs, combined with the heart, fill most of the space in the squirrel's upper body, called the thoracic cavity. The highly muscular diaphragm controls the movement and volume of air that passes to and from the lungs. As the diaphragm contracts and forces the rib cage to rise, negative pressure is created in the thoracic cavity, which makes the lungs expand, drawing in air.

CONNECTIONS

COMPARE the squirrel's heart with that of a much larger animal such as a HIPPOPOTAMUS. Despite the size difference between the animals, the basic structure is the same because they are both mammals.

COMPARE the squirrel's lungs with the gills of a GULPER EEL. Both organs allow respiratory gases to pass in and out of the blood to the surounding medium.

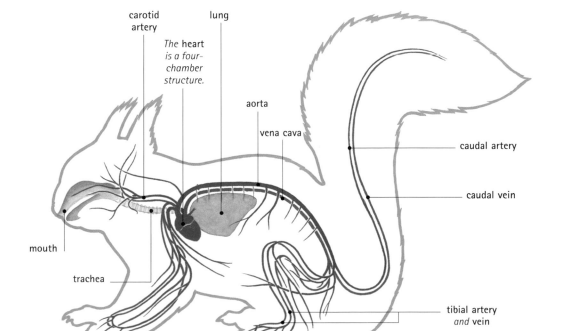

carotid artery
lung
The heart is a four-chamber structure.
aorta
vena cava
caudal artery
caudal vein
mouth
trachea
tibial artery and vein

◀ Eurasian red squirrel
The muscular heart pumps blood to the lungs, where it picks up oxygen from inhaled air. The oxygenated blood returns to the heart, where it is then pumped around the rest of the body to supply cells with oxygen for respiration.

Digestive and excretory systems

CONNECTIONS

COMPARE the function of a squirrel's large cecum with the short vestigial appendix of a *HUMAN*. They evolved from the same structure.

COMPARE the structure of a squirrel's digestive system with that of a carnivore such as a *WEASEL*. Meat is easier to digest than plant matter, and so a weasel's digestive tract is much shorter than that of a squirrel.

Squirrels are mainly herbivorous, so their digestive system is equipped to break down the tough cellulose found in plant matter. In the mouth, food is chewed and ground into a soft lump called a bolus and then swallowed. The bolus passes along the esophagus, which is lined with mucous membranes that allow the food to pass smoothly down into the stomach. The food is broken up and softened in the stomach by muscular action and enzymes, including pepsin. This enzyme requires an acidic environment to be effective and is activated by hydrochloric acid produced by cells in the stomach wall. The food is prevented from leaving the stomach during this process by two circular muscles, called sphincters, which are located at either end of the stomach.

From the stomach, the food passes through the pyloric sphincter and into the small intestine. Most absorption (movement of food molecules from the intestine into the blood) occurs when the food enters the large intestine. A pouch connected to the large intestine, called

the cecum, contains a multitude of bacteria. These microorganisms specialize in breaking down large cellulose molecules into simple sugars and starches. The squirrel is therefore far more efficient at digesting plant carbohydrates than animal proteins. The cecum also absorbs most of the water produced by digestive secretions, enabling it to be used elsewhere within the body. There is some variation in digestive abilities among species. For example, the robust digestive tract of eastern gray squirrels can cope with large quantities of chemicals called tannins, which occur in acorns, far more effectively than the digestive tract of red squirrels.

Food and feeding behaviors

Tree squirrels eat a wide variety of foods, including nuts, fruits, seeds, buds, catkins, sap, and even fungi and lichens. Ground squirrels forage mainly on grasses, roots, flowers, and bulbs. Many squirrels are opportunistic feeders and will supplement their diet with protein

▶ **Eurasian red squirrel**
An important feature of the squirrel's digestive system is the cecum. This pouch in the large intestine contains bacteria that break down the tough plant protein cellulose.

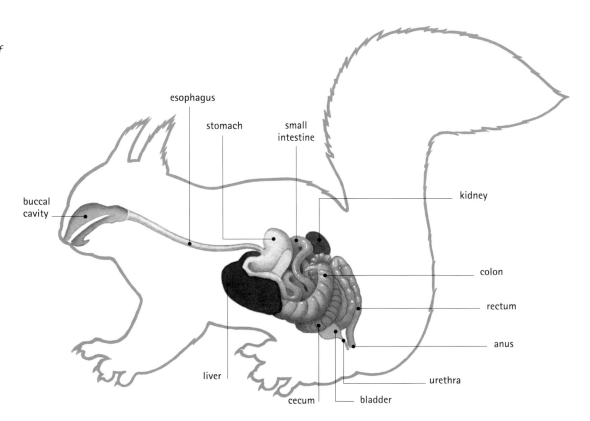

when available, from sources such as insects; small vertebrates, including frogs; and birds' eggs and nestlings. Both red and gray squirrels have often been observed chewing on animal bones, antlers, and even turtle shells—perhaps to supplement their calcium-poor diet with calcium and other essential minerals. Squirrels usually gain sufficient water from their regular food, but they may sometimes need to drink from puddles or pools during hot summer weather or in winter when food is scarce.

Storing food

Foraging and feeding may take up to 80 percent of a squirrel's active time, and in northern species this time increases markedly during the fall because the animal increases food intake to build up body fat and thicken its fur for winter. Many species store food in caches by burying nuts just below the topsoil and hiding pinecones or seeds in burrows.

Fungi are stored as single pieces higher in trees. Caches usually contain one to four items of food, and buried seeds are often not found again and so may germinate. That is why—despite their reputation as a forest pest—squirrels play a key role in the dispersal and regeneration of trees such as oak and beech.

▲ *An Arizona gray squirrel gnaws fruit. Squirrels eat a variety of foods, including nuts, fruits, buds, and fungi.*

CLOSE-UP

Feeding on fungus

Squirrels are able to eat certain fungi that contain highly poisonous amanita toxins. Their stomachs are lined with a layer of mucus made up of glycoproteins that neutralize the toxins and allow the squirrel to digest the fungi. The glycoproteins bond with the toxic element and make it indigestible, so instead of entering the bloodstream the toxins are excreted harmlessly. In this way, squirrels are able to feast on mushrooms that would poison or even kill other animals.

Reproductive system

COMPARE the developmental stages of a newborn squirrel with a newborn **GIRAFFE**. Squirrels are born virually helpless. A newborn giraffe, however, is highly developed.

COMPARE the large litter size of a squirrel with the small litters of an ape such as a **CHIMPANZEE** or a **HUMAN**. Ape babies are born large so there is not enough room in the uterus for several young.

CONNECTIONS

The reproductive anatomy of squirrels is very similar to that of other mammals. There is little external difference between the sexes, although males have a wider space—about 0.4 inch (1 cm)—between their genital and anal openings. A male's testes become swollen in the breeding season, and their color may darken, possibly owing to staining by urine. Females have a Y-shape reproductive tract and six pairs of nipples, which deliver the milk from mammary glands for suckling young. The milk is rich in proteins and fats, and nourishes the newborn pups throughout their early stages of development. Lactation usually lasts for up to nine weeks after the young are born.

Breeding and birth

Most Eurasian red squirrels become sexually mature at about 11 months old, and both sexes may be polygamous (they have numerous mates), particularly males. The female is in estrus for just one day of her cycle, and at this time her urine and vaginal secretions contain

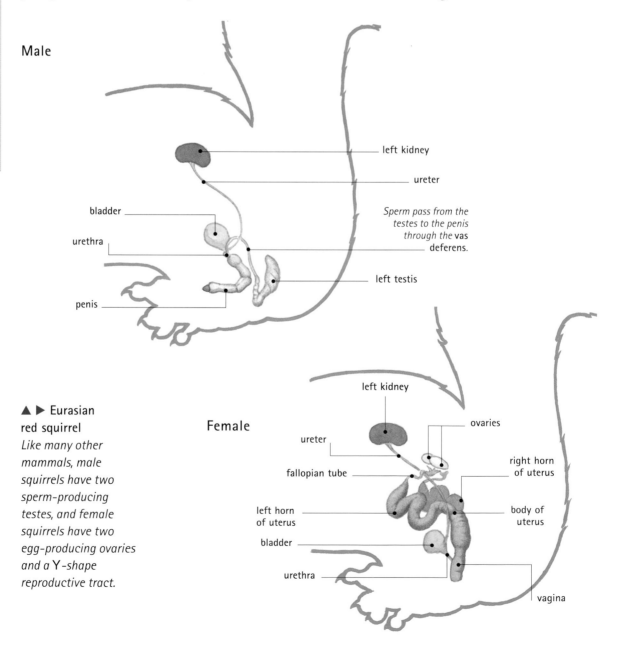

Male

left kidney

ureter

Sperm pass from the testes to the penis through the vas deferens.

bladder

urethra

left testis

penis

▲ ▶ Eurasian red squirrel
Like many other mammals, male squirrels have two sperm-producing testes, and female squirrels have two egg-producing ovaries and a Y-shape reproductive tract.

Female

left kidney

ovaries

ureter

right horn of uterus

fallopian tube

left horn of uterus

body of uterus

bladder

urethra

vagina

The development of young squirrels

Like all squirrels, Eurasian red squirrels are born blind, deaf, and hairless and are completely dependent on their mother for the first seven weeks of life. After eight or nine days, their skin darkens with pigmentation and the first hairs begin to sprout. By 21 days, the entire body is covered with fur and the lower incisor teeth have appeared. The upper incisors follow at 31 to 42 days. The pups' eyes open by the time they are 30 days old, and they are able to hear at between 28 to 35 days. The mother uses her tongue to stimulate the pups to urinate and defecate until about the seventh week, when they begin to eat solid foods. By this time, they are able to begin leaving the nest for short periods; by the eighth week they are fully weaned, although the female may continue to protect them for a few extra weeks. They finally gain a full coat of adult fur at three to four months old and then disperse.

◀ Like red squirrels, gray squirrels spend the first few weeks of life in the dray blind, deaf, and entirely dependent on their mother.

chemical messages that signal to males that she is ready to mate. The female is often pursued relentlessly by several suitors, who may fight for the chance to mate with her. Courtship is extremely brief: the male mounts the female, and immediately after mating is completed the pair separate. Fertilization occurs internally, and the fertilized eggs (zygotes) become implanted in the wall of the uterus. Each fetus, like those of most mammals, attaches to a placenta and has fetal membranes through which it receives nutrition and expels waste. The female gives birth at about 38 days and tends the young alone, often remaining in the nest, or dray, for long periods when the pups are very young. She may sometimes transport the pups in her mouth to a new dray and usually covers them with nesting material if she leaves to forage.

Many species of squirrels, such as rock squirrels and the northern flying squirrels, have a single litter each year in spring. However, several species, such as chipmunks and eastern gray and Eurasian red squirrels, have two litters annually. They usually breed from mid-January to April and again from July to September, although this may depend on availability of food and other environmental conditions. The first breeding in spring may be delayed or avoided if there are insufficient food stocks available, but equally the breeding season may be prolonged if there is a good crop of seeds and mild weather. Litter size is usually one to six pups, but up to 11 have been recorded, although the larger species usually have smaller litters.

STEVEN SWABY

FURTHER READING AND RESEARCH
Macdonald, David. 2006. *The Encyclopedia of Mammals.* Facts On File: New York.
Nowak, Ronald M. 1999. *Walker's Mammals of the World.* Johns Hopkins University Press: Baltimore, MD.

Starfish

PHYLUM: Echinodermata CLASS: Asteroidea
ORDER: Forcipulatida and six others

The 1,500 living species of starfish, or sea stars, are based on a body plan of a number of arms emerging from a central disk. Starfish are marine invertebrates, and most prey on other invertebrates.

Anatomy and taxonomy

Scientists categorize all organisms into taxonomic groups based partly on anatomical features. Starfish are asteroid (star-shape) echinoderms in the class Asteroidea.

● **Animals** Starfish, like other animals, are multicellular and obtain their food by consuming other organisms. Animals differ from most other multicellular life-forms, such as plants, in their ability to move from one place to another (usually using muscles). They generally react rapidly to touch, light, and other stimuli.

● **Metazoans** In metazoans the cells are combined in groups with cells of similar function to form tissues, such as muscle or nervous tissue. These tissues are often arranged into a structure such as an eye, stomach, or liver, which has a recognizable function. Metazoans range from simple cnidarians to complex invertebrates, such as octopuses and crabs, and vertebrates, such as fish.

● **Invertebrates** Along with other invertebrates, starfish lack a backbone (spine, or vertebral column), a feature that is found in vertebrates.

● **Echinoderms** There are about 7,000 different living species in the phylum Echinodermata (from the Greek *echinos*, meaning "spiny"; and *derma*, meaning "skin"). They are named for the elements of calcium carbonate (called ossicles) that are embedded in their skin and connected to one another by ligaments made of collagen, a type of protein. Each calcareous element is made from a form of

▼ Starfish make up the class Asteroidea, which in turn are echinoderms—marine invertebrates (animals without a backbone) with calcium-containing granules in their skin. Other echinoderms include sand dollars and sea cucumbers.

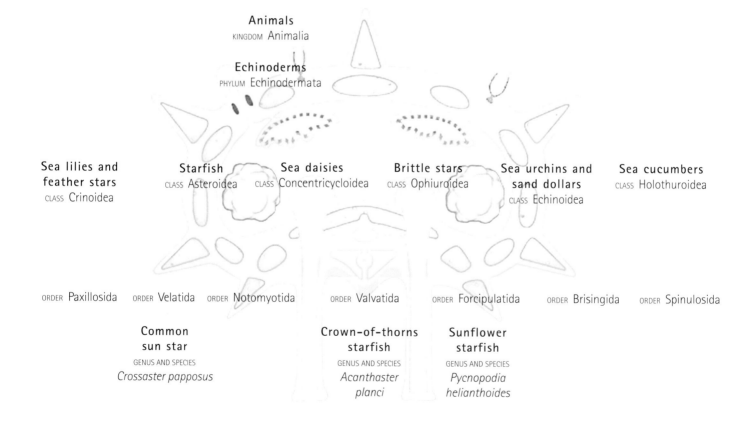

Animals
KINGDOM Animalia

Echinoderms
PHYLUM Echinodermata

Sea lilies and feather stars
CLASS Crinoidea

Starfish
CLASS Asteroidea

Sea daisies
CLASS Concentricycloidea

Brittle stars
CLASS Ophiuroidea

Sea urchins and sand dollars
CLASS Echinoidea

Sea cucumbers
CLASS Holothuroidea

ORDER Paxillosida   ORDER Velatida   ORDER Notomyotida   ORDER Valvatida   ORDER Forcipulatida   ORDER Brisingida   ORDER Spinulosida

Common sun star
GENUS AND SPECIES
Crossaster papposus

Crown-of-thorns starfish
GENUS AND SPECIES
Acanthaster planci

Sunflower starfish
GENUS AND SPECIES
Pycnopodia helianthoides

calcium carbonate called stereom, which is unique to echinoderms. This arrangement of stereom elements forms a true endoskeleton (inner skeleton) embedded in layers of living cells. Echinoderms range in size from minute sea urchins less than 0.2 inch (0.5 cm) across, to starfish such as the sunflower sea star, which is up to 39 inches (1 m) across.

Like the much simpler cnidarians and comb jellies, echinoderms are radially symmetrical: their body is made up of more or less identical radiating, repeated elements arranged around a central axis. However, their larvae are bilaterally symmetrical: the left half is a mirror image of the right half. Unlike cnidarians, most adult echinoderms have a body arranged into five similar regions around a circular shape, a condition called pentaradial symmetry. A unique feature of echinoderms is their water–vascular system.

● **Brittle stars** The 2,000 or so species of brittle stars belong to the class Ophiuroidea, the largest group of echinoderms. Brittle stars have five thin, flexible, brittle arms. Brittle stars use their tube feet for plucking animals or small particles from the seabed and passing them, from foot to foot, to the mouth. Brittle stars can travel by the snakelike movement of their arms.

● **Sea lilies and feather stars** There are about 700 species of sea lilies and feather stars, members of the class Crinoidea. They are similar in appearance to an upside-down brittle star, with the mouth and furrows called ambulacral grooves directed upward rather than downward.

▲ This starfish shows the pentaradial symmetry typical of most adult echinoderms: it has five arms radiating from a central disk.

● **Sea urchins and sand dollars** The 1,200 or so species of sea urchins (class Echinoidea) have an endoskeleton that forms a rigid test (a shell-like structure) armed with movable spines and pedicellariae (minute pincerlike extensions). Some sea urchins burrow in sand and feed on buried deposits.

● **Sea cucumbers** Sea cucumbers (class Holothuroidea) have a wormlike shape. They lack spines, and their body wall is warty and contains numerous microscopic ossicles. Sea cucumbers travel along the seabed on rows of tube feet and also by extending and contracting their body.

● **Sea daisies** The three species of sea daisies have been discovered since the mid-1980s. These creatures are disk-shaped, lack arms, and have a ring of spines at their outer edge. Sea daisies are tiny and have a poorly developed gut.

● **Starfish** Starfish, or sea stars (class Asteroidea), are the most familiar type of echinoderms. Many species have five arms emerging from a central disk, but some have many more arms, including the sunflower sea star, which typically has about 24. Starfish move along on hundreds of tube feet in the ambulacral grooves under each arm. Starfish also bend their arms to explore the environment and capture prey. Most starfish prey on other invertebrates, including snails, bivalve mollusks, and other echinoderms.

FEATURED SYSTEMS

EXTERNAL ANATOMY Starfish have a spiny skin. Small calcium carbonate elements, called ossicles, connected by a network of ligaments, form an inner skeleton just beneath the body surface. *See pages 1226–1228.*

INTERNAL ANATOMY A hydrostatic (water-pressure) system connected to tube feet enables the starfish to "walk" on any type of surface. *See pages 1229–1331.*

NERVOUS SYSTEM Despite having a decentralized nervous system without a recognizable brain, starfish show relatively complex behavior. *See page 1232.*

CIRCULATORY AND RESPIRATORY SYSTEMS The starfish has three kinds of circulatory system, only one of which contains blood. *See page 1233.*

DIGESTIVE AND EXCRETORY SYSTEMS The digestive system is simple. Many starfish can push part of their stomach out through the mouth to digest prey items outside the body. *See pages 1234–1235.*

REPRODUCTIVE SYSTEM During development, starfish undergo a remarkable metamorphosis from a bilaterally symmetrical larva to a radially symmetrical adult. *See pages 1236–1237.*

External anatomy

The sunflower starfish is a very large starfish with a broad central disk and typically about 24 flexible arms when adult, although young begin life with five and quickly add a sixth. This species can be yellow, but it comes in a variety of other shades, including orange, red, brown, pink, and purple. It is named for its superficial resemblance to a sunflower.

When stranded on the shore or held in the hand, the soft, flabby body makes the sunflower starfish appear relatively helpless. However, when submerged in water the animal is able to move about on the seabed at about 5 feet (1.5 m) a minute, which is fast for a starfish. It is a voracious predator of small bottom-living animals.

An animal without a head

In common with other radially symmetrical animals, most echinoderms do not have an obvious front (anterior) and back (posterior) end. (Sea cucumbers are an exception.) Instead, the surface where the mouth is located is called the oral surface, and the opposite side (usually on top) is called the aboral surface. The oral surface of a starfish has a mouth at the center.

39 inches
(100 cm)

▼ **Sunflower starfish**
The sunflower starfish has up to 24 arms and can move at 5 feet (1.5 m) per minute. Fully grown, it is up to 39 inches (1 m) from arm tip to arm tip. This species may be pink, purple, brown, red, or yellow.

*The **oral disk** is the central part of the body and contains all the main organs, including intestines and a stomach.*

*The **madreporite** is the entrance to the water-vascular system.*

*Many **papulae** cover the surface of the arms.*

Tube feet *are used for locomotion and respiration.*

Arms *grow from a central disk. A young sunflower starfish has only six arms, but it grows more as it matures until it has between 16 and 24 arms. If an arm is severed, the starfish can grow a replacement from its oral disk.*

The upper (aboral) surface is covered by **armored skin**.

Running along the underside of each arm is a furrow, called the ambulacral groove, containing two or four rows of tube feet, depending on the species. Along the edges of the groove is a row of short spines that can be folded over the furrow to protect the tube feet. Concentrated at the tip of each arm are sensory structures: touch-sensitive and chemical-detecting tube feet and a pigment spot that probably serves as a simple eye. Because starfish can move in any direction, they sense the environment around them from the extreme tips of the arms. These tips are the first parts of the starfish's body to encounter new surroundings.

The aboral surface of a starfish has fewer obvious features than the underside. The madreporite, a buttonlike structure, sits slightly off-center on the central disk and between two arms. The madreporite is the entrance to a starfish's water-vascular system. This system, unique to echinoderms, delivers fluid to structures used in locomotion (the tube feet). Papulae (singular, papula), miniature extensions of the body wall, allow gas exchange between the water outside and the fluid inside the body cavity. Thus, papulae are called dermal gills. The anus, the exit of the digestive system, also lies on the aboral surface, when it is present at all.

On both the oral and aboral surfaces some ossicles in the skin form spines, which provide both support and some protection against attack by predators. The skin of both surfaces is armed with pedicellariae (singular, pedicellaria), which are like miniature pincers or scissors. There are several kinds of pedicellariae, with some types occurring on only one or the other surface of the starfish. Protective types of pedicellariae grasp any animal or other object that touches the skin, and they form a means of defense. In some starfish species, such as *Stylasterias forreri*, groups of pedicellariae are an important means by which the starfish captures food.

▶ TUBE FEET

Tube feet are not only used for locomotion. They are part of the starfish's water-vascular system and work hydraulically: water is forced into them, and they elongate. Tube feet also secrete chemicals that help the feet attach to and then release objects.

CLOSE-UP

Jaws in the skin

The surface of a starfish is scattered with jawlike structures called pedicellariae. In the sunflower sea star, a single pedicellaria consists of a short, fleshy stalk topped with pincerlike, or scissorlike, jaws supported by small, hard plates called ossicles. In most cases, pedicellariae defend a starfish against attack by small creatures and against small larvae that could grow on the starfish's body. When pedicellariae were first discovered by a researcher in the 18th century, he thought the structures were parasites living on the surface of the animal. Now it is known that the structures are produced by the starfish and other echinoderms themselves. Pedicellariae have their own nervous and muscle systems. Most puzzling of all, the pedicellariae do not seem to be supplied with food from the gut or the water-vascular system. They may absorb chemicals for their own needs from the environment.

▶ PEDICELLARIAE

The aboral surface of a starfish is covered with pedicellariae and spines.

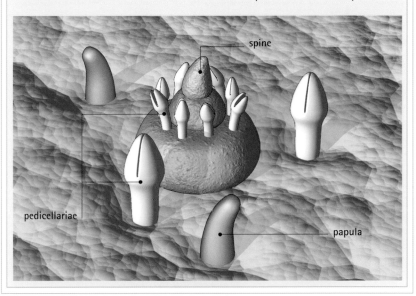

spine

pedicellariae

papula

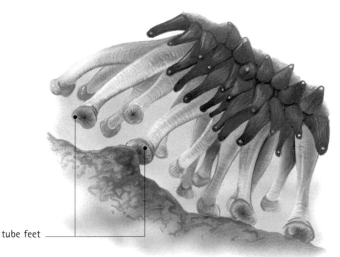

tube feet

Echinoderm body shape

Present-day echinoderms probably evolved from a common ancestor in the Cambrian period more than 500 million years ago. That was a time when invertebrates were undergoing an explosion in diversity as many new habitats became available in shallow water. Early ancestors looked little like modern forms. Although they had an endoskeleton of ossicles, many were not radially symmetrical. They were probably more or less stationary and filtered the water for suspended food. Modern echinoderms are variations on the pentaradial theme. All have ambulacral grooves with rows of tube feet, showing how the different forms are variations on a common design. Brittle stars (ophiuroids) are essentially spindly versions of starfish but with a clearly demarcated central disk. Sea urchins (echinoids) can be imagined as starfish with the five arms folded back toward the aboral (upper) surface and joined up to create a roughly spherical shape. Sea cucumbers (holothuroids) have a pentaradial arrangement of tentacles around the mouth. Sea lilies and feather stars (crinoids) are like stalked or clawed versions of upside-down brittle stars.

▼ *The slate pencil urchin is a relative of starfish. Pencil urchins use their many flattened spines to move around on the shallow seabed. They often hide from predators in crevices in coral reefs.*

The outer body wall

A starfish's body surface is covered by an outer layer, the epidermis (skin), made up of several types of cells. Most of the cells have tiny hairlike structures called cilia, and some of these cells are sensory, responding to touch or to chemicals in the environment. Some cells lack cilia and are secretory, releasing sticky mucus that helps protect the body surface. Particles, bits of debris, and small organisms become trapped in the mucus and are wafted by moving cilia toward the mouth, where they are either eaten or rejected and left behind in the wake of the starfish. Near the base of the epidermis lies an interconnecting network called the nerve plexus, which contains nerve fibers (the elongated parts of nerve cells). The fibers transmit messages, as electrical impulses, to and from the epidermis, allowing the animal to sense its surroundings and respond by coordinating the activity of structures, such as tube feet, on the body surface.

Beneath the outermost layer of the starfish's body surface, the epidermis, lies a thick layer of connective tissue, the dermis, which contains the hard parts of a starfish's skeleton, the ossicles. The ossicles are made mostly of a calcium carbonate mineral called calcite, which includes some magnesium carbonate. Each ossicle is made of numerous tiny crystals that are aligned to create a polycrystalline unit that acts like a single crystal of calcite. Ossicles come in a variety of shapes, including rods, crosses, and plates. The ossicles are connected to one another by ligaments made of the protein collagen. Ossicles grow inside the dermis and are secreted by the cells that surround them.

The connective tissue of the dermis is remarkable in being able to suddenly change in consistency, from a rigid gel into a runny fluid. Beneath the dermis lie sheets of circular and longitudinal muscle fibers that work against each other to bend a starfish's arms.

The ability to regenerate

Echinoderms, including starfish, have an astonishing capacity to regenerate body parts that are damaged or removed. Most species of starfish can regenerate themselves completely from a single arm and part of the central disk. The wounds seal themselves, and at the sites of injury new tissue develops and grows. Brittle stars readily shed their arms as a strategy to keep an attacking predator occupied. The detached, writhing arm holds the predator's attention while the brittle star escapes. It soon regrows its lost arm.

Internal anatomy

Echinoderms possess a unique system of fluid-filled cavities and tubes that play a central role in movement. During the animal's development, this water-vascular system originates from one of the coeloms (fluid-filled cavities in the middle of the second of the body's three layers that form during development). The water-vascular system is lined with ciliated cells. The cilia propel coelomic fluid around the body through the system of canals.

Seawater system

The water-vascular system is unique among complex animals in being supplied by seawater direct from the surroundings. In complex marine animals other than echinoderms, fluid either is swallowed or passes by diffusion across the body wall before it enters internal systems. In echinoderms, seawater enters the water-vascular system through a structure called the madreporite, which contains a system of pores. Cilia guard the pores and help prevent unwanted particles entering or blocking them. Seawater is drawn through the madreporite by cilia beating within the water-vascular system. Seawater flows slowly into a structure called the stone canal, which descends vertically from the

CLOSE-UP

Digging feet

Many species of starfish that live on soft seabeds covered with sand or mud have tube feet geared for digging rather than walking. The tube feet do not end in suckers but are pointed, and they can be driven between grains of sand. At its base each tube foot has two muscular sacs called ampullae rather than one (as in other types of starfish), and these provide more power for digging. Some starfish use their tube feet for excavating burrows and lining the walls of the burrows with mucus.

COMPARE the suckers on the tube feet of a starfish with those on the arms of an *OCTOPUS*. Most starfish have a single sucker at the end of each tube foot, whereas two rows of suckers run down the length of an octopus's arm. They are used for touching and for holding on to prey.

CONNECTIONS

▼ *Starfish feel their way around using their tube feet, part of the hydrostatic water-vascular system.*

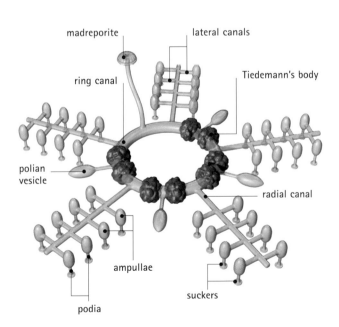

▲ WATER-VASCULAR SYSTEM
The water-vascular system is a network of water-filled canals. The canals extend into muscular extensions called tube feet, which are extended and withdrawn by a hydrostatic (water-pressure) system.

(labels: madreporite, lateral canals, ring canal, Tiedemann's body, polian vesicle, radial canal, ampullae, suckers, podia)

Tube feet on the march

When a starfish is moving sideways over the seabed, each tube foot performs a stepping motion slightly out of phase with the one alongside. The tube feet working together in their hundreds gradually move that part of the animal forward.

Each "step" of a tube foot begins with muscles contracting in the wall of the ampulla, a muscular sac at the base of the tube foot; and the closing of the valve in the adjacent lateral canal, which supplies the ampulla with fluid. These actions force fluid into the tube foot, making it extend. At the same time, longitudinal muscles on one side of the wall of the tube foot contract slightly, causing it to bend in the forward direction. The sucker at the tip of the tube foot then adheres to a surface, partly by chemical adhesion and partly because muscles in the center of the sucker contract, turning the sucker into a suction cup. Some of the longitudinal muscles now contract, causing the tube foot to bend backward. The shift in position of the tube foot drags that part of the animal forward. Longitudinal muscles in the tube foot now contract more generally, causing the tube foot to shorten and fluid to return to the ampulla. The sucker then detaches, and the tube foot is now ready to take its next step.

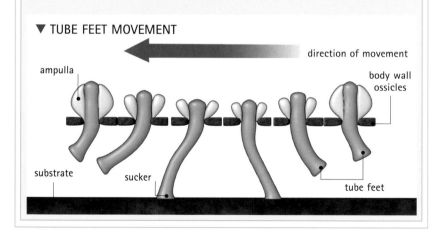

▼ TUBE FEET MOVEMENT

direction of movement

ampulla

body wall ossicles

substrate

sucker

tube feet

madreporite. The stone canal is named for the calcium carbonate deposits embedded in its walls. After seawater flows down the stone canal, it reaches a canal encircling the esophagus. From this canal radiate five or more channels called radial canals. One radial canal supplies each of a starfish's arms. Typically, two pairs of folded pouches called Tiedemann's bodies (named for the scientist who discovered them) lie in the ring canal between each radial canal. Tiedemann's bodies contain cells called phagocytes, which remove bacteria and other harmful invaders from the incoming water. This protective function is similar to that performed by the lymph nodes in vertebrates. Some

starfish have muscular sacs arranged around the ring canal. These may help regulate the water pressure within the water–vascular system.

Lateral canals

From the radial canal, side branches, called lateral canals, extend into each arm. Each lateral canal ends in a muscular sac, or bulb, called an ampulla, that leads into a tube foot. A valve in each lateral canal can temporarily close off the ampulla and tube foot from the rest of the water–vascular system. Beating cilia in the lining of the water–vascular system circulate fluid. Contractions of muscles working with sets of valves raise the water pressure in parts of the system. These pressure changes allow the animal to extend and contract its tube feet, enabling it to "walk" on

▲ If a starfish gets turned upside down, it can right itself by folding one or more arms back under its body. The tube feet on these arms attach to the sea bottom, and once they start "walking" they drag the rest of the body over.

Stuck on you

The suckers on tube feet stick to the seabed by a combination of chemical attraction and suction. The sucker releases a type of glue that sticks it to the seabed. This glue is similar to the type of adhesive produced by flatworms and snails. When the starfish's tube foot releases another chemical, it causes the glue to lose its stickiness and fall apart, allowing the tube foot to be released.

▼ STALKED PEDICELLARIA
Common starfish
Starfish use pedicellariae for protection against animals that may settle on their arms or on the oral disk. There are two kinds of pedicellariae: stalked and sessile. A stalked pedicellaria has a jawlike apparatus on a fleshy stalk. Pairs of muscles open and close the jaws.

the seabed. The tube feet, using suction and chemical "glues," also serve to anchor the animal to the seabed or almost any other object. Starfish have no trouble using their tube feet to climb up the vertical glass in aquariums. Starfish also use their tube feet to grasp and manipulate items of prey when feeding.

Arm muscles
Two sheets of muscles extend under the body wall of each arm. The outer sheet is circular muscle (with muscle fibers arranged in rings around the arm), and the inner sheet is longitudinal (muscle fibers arranged parallel to the arm's length). The circular muscles are used to flatten the arm, and the longitudinals raise and lower it, or swing it from side to side. In addition, fine control of the contraction and relaxation of the two muscle sheets can make an arm bend in any direction. The flexible arms, combined with the array of tube feet on their underside, enable a starfish to move about on almost any type of surface, whether rough, smooth, hard, or soft.

▼ CROSS SECTION THROUGH ARM
The main muscles and features of the water-vascular and digestive systems are shown. Ambulacral muscles work the water-vascular system, enabling the starfish to extend and contract its tube feet. Arm muscles allow the arms to bend, lengthen, or shorten.

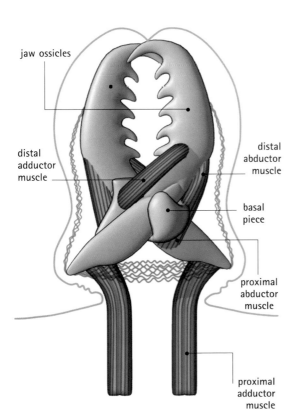

jaw ossicles

distal adductor muscle

distal abductor muscle

basal piece

proximal abductor muscle

proximal adductor muscle

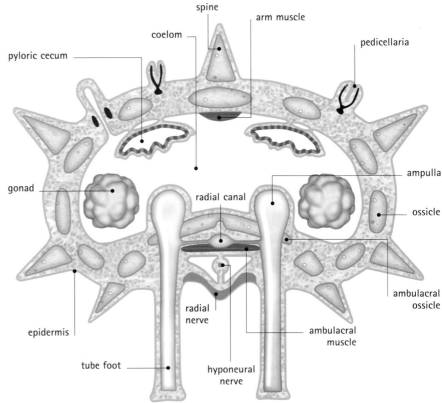

spine

coelom

arm muscle

pyloric cecum

pedicellaria

gonad

ampulla

radial canal

ossicle

epidermis

ambulacral ossicle

radial nerve

ambulacral muscle

tube foot

hyponeural nerve

Nervous system

CONNECTIONS

COMPARE the nervous system of a starfish with that of a *JELLYFISH*. Both types of animals are invertebrates with no brain and a nervous system arranged on a radial plan.

Starfish and other echinoderms have a relatively simple nervous system based on a radial plan, similar to that found in cnidarians such as jellyfish. Because starfish and most other echinoderms do not have a recognizable head, sensory structures are scattered about the body rather than concentrated in one region. Although starfish are capable of quite complex behavior, such as when prying open clam shells or righting their body when overturned, their nervous system is highly decentralized. There is no recognizable brain.

The main part of a starfish's nervous system is a ring of nervous tissue, the circumoral nerve ring, which encircles the esophagus and connects to a radial nerve running along each arm. Where the nerve ring and radial nerve meet, a collection of nerve cells helps coordinate activity in the adjacent arm. From each radial nerve emerge smaller branches, called abradial nerves, that connect to muscles and sensory structures in the ampullae, tube feet, and elsewhere.

The nervous system in each arm is really two nerve networks. An oral system lies closest to the underside of the starfish, and running alongside and just above lies the deep oral system. The oral system is mainly sensory, carrying information from sensory cells in the epidermis and distributing this information to other parts of the body. The deep oral system carries mainly motor information—instructions to contract muscles and to stimulate other responding organs. The nerve plexus—a network of nerve fibers at the base of the epidermis—contains fibers from both motor and sensory nerve cells.

Starfish and other echinoderms do not have complex sense organs, but they have many sensory cells. Scattered over the surface of its body, and especially on the tube feet, a typical starfish has more than 70,000 sensory cells. These respond variously to touch, dissolved chemicals, light, and water currents.

At the tip of each starfish arm is a tentacle-like structure, which probably evolved from a tube foot. This structure contains 80 to 200 cup-shape, light-sensitive structures called ocelli. These ocelli together make up an eyespot. Although this structure is not as sophisticated as an eye, it helps the starfish orient itself relative to the direction of incoming light. Most starfish tend to move toward the light rather than away from it.

▶ **NERVES OF OUTER BODY WALL**

The nerve plexus runs just beneath the outer layer of cells, the epidermis. Some of the cells of the epidermis are sensory cells, which connect directly with the nerve plexus.

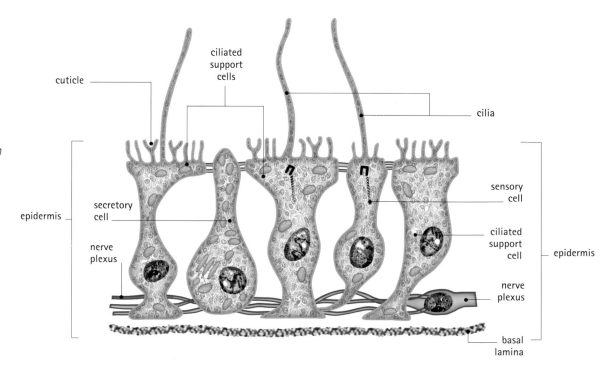

cuticle
ciliated support cells
cilia
epidermis
secretory cell
sensory cell
nerve plexus
ciliated support cell
epidermis
nerve plexus
basal lamina

Circulatory and respiratory systems

Starfish have three kinds of circulatory systems. The first, the water-vascular system, connects to ampullae and tube feet, providing these structures with nutrients absorbed from the gut. The second system contains coelomic fluid and is divided into three parts: one that supplies the nervous system; a second that supplies the gonads (testes and ovaries); and a third, more extensive, part that bathes many structures of the central disk and arms, supplying them with nutrients and oxygen and removing waste products. The coelomic fluid is circulated within the whole system by the beating of cilia on the surface of cells that line the canals and cavities.

The third system is a blood-vascular system. It contains colorless blood that is pumped around the body by a simple heart that lies just below and to one side of the madreporite. Blood vessels carry nutrient-rich blood from the gut to certain parts of the body such as the gonads. Some structures near the heart—the gastric hemal tufts and the axial gland—may serve as kidneys that filter waste substances from the blood. Their role is being investigated by researchers.

Gas exchange

Starfish gain oxygen from their surroundings and get rid of carbon dioxide by diffusion across many parts of the body surface. This exchange of gases occurs especially across the surface of tube feet and outfoldings called papulae, or dermal gills. Papulae are lined with cells of the epidermis on the outside and a thin layer of tissue called the peritoneum on the inside, and they contain coelomic fluid. The beating of cilia on the epidermis and peritoneum creates currents of seawater and coelomic fluid that move in opposite directions (a countercurrent flow). This arrangement ensures a steep concentration gradient (a large difference in concentration) between dissolved gases inside the starfish's coelom and dissolved gases in the surrounding seawater. Steep concentration gradients maximize the rate of gas exchange.

COMPARE the countercurrent flow across the papulae of a starfish with that in the gills of a *TROUT*. In both, a countercurrent flow maximizes the amount of oxygen picked up from the water.

CONNECTIONS

▲ SECTION THROUGH PAPULA

Starfish gain oxygen from seawater by diffusion across many parts of the body surface. Much diffusion occurs through the surface of structures called papulae. There, the beating of cilia creates currents of seawater and coelomic fluid that flow in opposite directions.

movement of seawater

cilia

epidermis

coelomic fluid

movement of coelomic fluid

peritoneum

cilia

ossicles

dermis

▼ COELOMIC AND BLOOD-VASCULAR SYSTEMS

A simple heart pumps colorless, nutrient-rich blood around the body in a series of blood vessels.

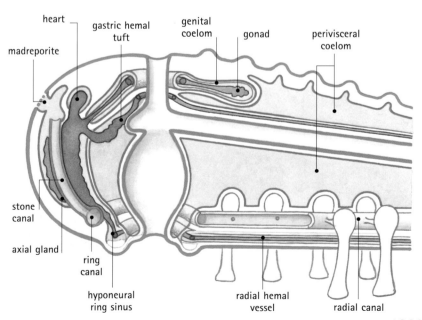

heart

gastric hemal tuft

genital coelom

gonad

perivisceral coelom

madreporite

stone canal

axial gland

ring canal

hyponeural ring sinus

radial hemal vessel

radial canal

Digestive and excretory systems

CONNECTIONS

COMPARE digestion in the starfish with that in a *TARANTULA*. A starfish can extend part of its stomach through its mouth and digest food before bringing it back into the body. A tarantula, however, vomits digestive juices onto food before ingesting the semidigested pulp.

Starfish are predators and scavengers that feast on a wide range of foods, which they find by smell and touch. Most types of starfish, including the sunflower starfish, can extend, or evert, the first part of the stomach, the cardiac stomach, through the mouth. This arrangement ensures that starfish can engulf and digest foods without having to swallow them first. The prey can be digested where they are found, and the products of digestion can then be carried back into the starfish's body. Some starfish pry open the shells of clams and mussels, push their stomach through the gap in the shell, and digest the victim still within its shell.

Simple gut

A starfish's gut is fairly simple in construction, with the mouth on the animal's underside leading directly into a short esophagus, which then enters a large stomach in the middle of the central disk. The stomach is divided into two halves: a cardiac stomach and a pyloric stomach. The breakdown of food begins in the cardiac stomach, where digestive enzymes released by the stomach wall begin to digest the item of food.

Digestion continues in a similar way in the pyloric stomach, but from this region two branches extend into each arm of the starfish. These extensions, called digestive glands or pyloric ceca, receive partially digested food from the stomach and digest it further, both within the chambers of the digestive glands and in the cells lining them. Partially digested

▶ **VERTICAL CROSS SECTION OF CENTRAL DISK AND PART OF ARM**
Some starfish have pockets of intestine called rectal glands, which recover nutrients that might otherwise be lost from the body.

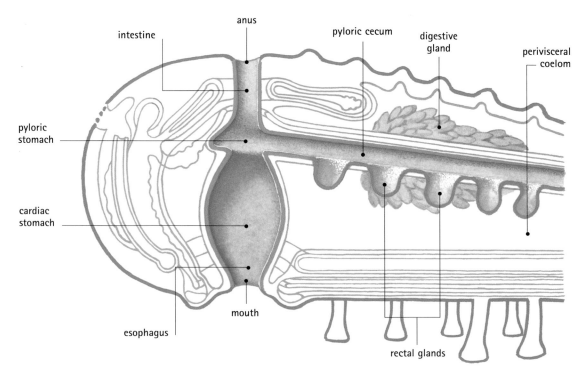

anus
intestine
pyloric cecum
digestive gland
perivisceral coelom
pyloric stomach
cardiac stomach
esophagus
mouth
rectal glands

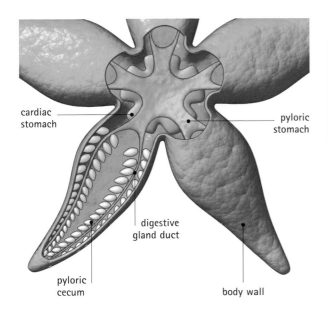

cardiac stomach

pyloric stomach

digestive gland duct

pyloric cecum

body wall

▲ VIEW FROM ABOVE
Part of the exoskeleton has been cut away to show the main features of the digestive system.

foods are routed through the digestive system by a combination of beating cilia and contracting muscles in the walls of the stomach and intestine. The final products of digestion pass out through the lining of the digestive glands and are transported around the body by the coelomic and blood-vascular systems to supply body cells with the nutrients they need. Any undigested waste passes out of the body from the pyloric stomach through a short intestine to the anus, which lies in the middle of the central disk on the aboral, or upper, surface. In some species lacking an anus,

PREDATOR AND PREY

The crown of thorns

The crown-of-thorns starfish of the tropical Pacific Ocean is a notorious consumer of coral polyps—the organisms that create coral reefs. The starfish everts its stomach onto a section of reef and digests the coral polyps where they lie. A crown of thorns can consume an area of coral of about its own size within a day. Huge populations of these starfish have devastated large areas of Australian coral since the 1980s. Until recently, scientists thought that human influences such as pollution or overfishing had killed off predators of starfish such as a large marine snail called the giant triton. Without these predators, starfish populations could increase to enormous numbers. However, recent fossil studies show that starfish "plagues" have been a regular feature over thousands of years.

the waste is emitted via the mouth itself. In some species of starfish, outpockets of the intestine called rectal glands recover some nutrients that might otherwise be lost through the anus.

Unlike crustaceans, mollusks, and other complex marine invertebrates, echinoderms in general do not have special excretory organs. Instead, nitrogen-containing wastes such as ammonia pass by diffusion across the body surface and into the surroundings. However, starfish can also pack their papulae with phagocytes that have engulfed potentially harmful organisms and waste substances. Starfish can pinch the papulae off at the base, so jettisoning the papulae and dumping waste-containing cells into their surroundings.

◀ *A starfish feeds on a bivalve mollusk. Some starfish are able to force apart the two shells of a bivalve to reach the edible parts inside. Starfish can push their cardiac stomach through the mouth to digest prey before swallowing it.*

Reproductive system

COMPARE fertilization in starfish with that in a **BULLFROG, GIANT CLAM, NEWT,** and **TROUT.** In all these animals, eggs are fertilized by sperm externally—outside the body in the water.

In most starfish, the sexes are separate. Males have sperm-producing organs called testes, and females have egg-producing organs called ovaries. Typically, two testes or ovaries are packed inside each starfish arm. When the breeding season approaches, the testes and ovaries swell with their sperm or eggs. Each testis or ovary has one or more tubes that connect to the body surface and exit through gonopores, which lie between the arms.

Males and females commonly come together in large groups to shed their sex cells into the surrounding water, where external fertilization takes place. Gathering in large numbers and releasing eggs and sperm at the same time will increase the likelihood of successful fertilization. The female starfish typically releases 1 million or more eggs, but many of these will go unfertilized or will be consumed by other animals soon after fertilization. Spawning usually occurs at night to reduce levels of predation.

Plankton

The fertilized eggs float up to the surface, where they form part of the plankton:

▶ **LARVAL DEVELOPMENT AND METAMORPHOSIS**

The diagrams show the development of a larva through its earliest bipinnarian and brachiolarian stages (1, 2, and 3), and then metamorphosis (4). At stage 5 the young starfish begins to develop in the posterior region of the old larva. At stage 6, the starfish attaches to the substrate and grows rapidly.

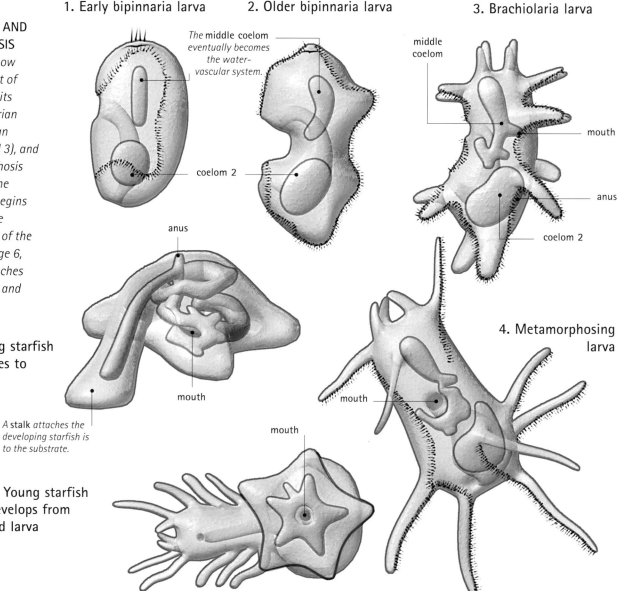

1. Early bipinnaria larva

2. Older bipinnaria larva

3. Brachiolaria larva

The middle coelom *eventually becomes the water-vascular system.*

coelom 2

anus

middle coelom

mouth

anus

coelom 2

6. Young starfish continues to develop

mouth

A stalk *attaches the developing starfish is to the substrate.*

5. Young starfish develops from old larva

4. Metamorphosing larva

mouth

mouth

mouth

organisms that drift with water currents. There, the eggs develop into larvae that swim by beating the rows of cilia on their surface. The larvae are bilaterally symmetrical (one side is the mirror image of the other), and they develop through two main stages, called bipinnaria and brachiolaria, until the final stage undergoes a remarkable transformation.

A miniature starfish develops inside the posterior end of the brachiolaria larva, while the anterior end develops three sticky arms for attachment. Eventually the larva descends to the seabed, and its front end attaches to the sediment and develops a stalk. This attachment process is the trigger for the starfish to develop rapidly. The larva breaks from the stalk, and the upper portion emerges from the posterior as a miniature version of the adult.

TREVOR DAY

FURTHER READING AND RESEARCH

Brusca, R. C., and G. J. Brusca. 2003. *Invertebrates.* 2nd ed. Sinauer Associates: Sunderland, MA.

Ruppert, E. E., R. S. Fox, and R. B. Barnes. 2004. *Invertebrate Zoology: A Functional Evolutionary Approach.* 7th ed. Brooks Cole Thomson: Belmont, CA.

CAS Echinoderm Webpage: www.calacademy.org/research/izg/echinoderm

EVOLUTION

Related to chordates?

Two main lines of evidence suggest that echinoderms are closely related to the direct ancestors of chordates, the group that includes all vertebrates. At a very early stage in development, when the fertilized egg is dividing to become a ball of cells, the cells of echinoderm and chordate embryos divide in a similar way, by spiral cleavage. Also, the first opening to appear in the ball of dividing cells marks the position of the anus. In all complex invertebrates other than echinoderms, embryos develop by another form of cell division, called radial cleavage, and the first opening to appear becomes the mouth. Biologists refer to chordates and echinoderms as deuterostomes (meaning "mouth second"), whereas other invertebrates are protostomes ("mouth first").

Another link between chordates and echinoderms is in the appearance of the larvae. The early-stage larvae of echinoderms and hemichordates, a primitive group of wormlike chordates, are almost identical. Thus echinoderms are probably the invertebrate group that is most closely related to chordates. Nevertheless, the two are distantly related at best. The line of descent to present-day chordates must have split off from the echinoderm line of evolution more than 400 million years ago.

▼ *Off the coast of California, the starfish* Asterias forbesi *releases a cloud of eggs or sperm into the ocean. This action is called spawning. Many of the sex cells will fertilize, or be fertilized by, sex cells from other starfish of the same species.*

Stingray

CLASS: Chondrichthyes ORDER: Myliobatiformes
FAMILY: Dasyatidae GENUS: *Dasyatis*

The southern stingray lives in the tropical and subtropical waters of the western Atlantic Ocean, including the Gulf of Mexico and the Caribbean Sea. Like many other rays, it lives in shallow water and prefers a sandy or silt–covered seabed. It feeds on a wide variety of bottom-living animals, including fish, crustaceans, and mollusks, which it senses mainly by smell, touch, and electrodetection.

Anatomy and taxonomy

Scientists categorize all organisms into taxonomic groups based largely on anatomical features. The southern stingray is one of 11 stingray species in the genus *Dasyatis*. Stingrays, along with closely related rays that do not have a stinger, belong to the order Myliobatiformes within the subclass Elasmobranchii, which includes all rays as well as sharks and skates.

● **Animals** These organisms are multicellular and gain their food supplies by consuming other organisms. Animals differ from other multicellular life-forms in their ability to move from one place to another (in most cases, using muscles). They generally react rapidly to touch, light, and other stimuli.

● **Chordates** At some time in its life cycle, a chordate has a stiff, dorsal (back) supporting rod called the notochord that runs all or most of the length of the body.

▶ *This family tree shows that stingrays are vertebrates (animals with a backbone) in the subclass Elasmobranchii, which also includes sharks, skates, and other rays. The southern stingray is in the family Dasyatidae and genus Dasyatis.*

● **Vertebrates** In living vertebrates, the notochord develops into a backbone (spine or vertebral column) made up of units called vertebrae. The vertebrate muscular system that moves the body consists primarily of muscles that are arranged in mirror image on either side of the backbone or notochord (bilateral symmetry about the skeletal axis).

● **Gnathostomes** These fish have jaws, unlike hagfish and lampreys (agnathans or jawless fish), which lack proper jaws. Gnathostomes have gills that open to the outside through slits, and fins that include those arranged in pairs, such as the pectoral (shoulder) fins.

● **Cartilaginous fish** Most fish have a skeleton made of bone. However, in cartilaginous fish, such as sharks, skates, rays, and chimeras, the skeleton is made of cartilage, a softer, more flexible material than bone. The skin is rough, like sandpaper, owing to the presence of enamel-coated toothlike scales called denticles. In addition, unlike bony fish, cartilaginous fish do not have a swim bladder.

Animals
KINGDOM Animalia

Vertebrates
SUBPHYLUM Vertebrata

Jawless fish
SUPERCLASS Agnatha

Gnathostomes
SUPERCLASS Gnathostomata

Bony fish
CLASS Osteichthyes

Cartilaginous fish
CLASS Chondrichthyes

Sharks, skates, and rays
SUBCLASS Elasmobranchii

Skates and rays
SUPERORDER Batoidea

Sharks
SUPERORDER Selachii

Guitarfish
ORDER Rhinobatiformes

Sawfish
ORDER Pristiformes

Stingrays
ORDER Myliobatiformes

Electric rays
ORDER Torpediniformes

Skates
ORDER Rajiformes

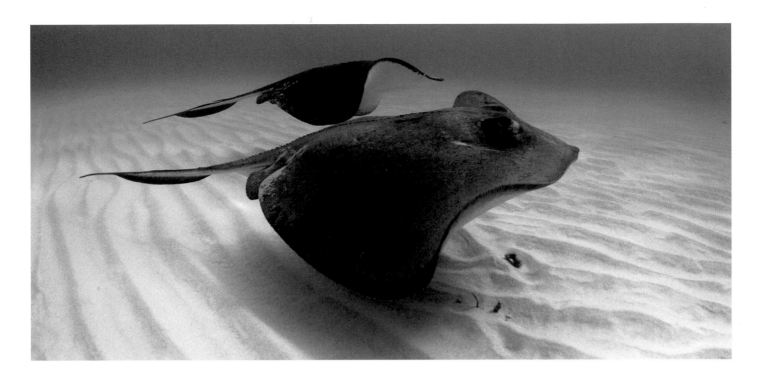

● **Sharks, rays, and skates** These fish belong to the largest group of cartilaginous fish, the elasmobranchs, which are members of the subclass Elasmobranchii. Most sharks are fusiform (torpedo-shaped) and are active swimmers that feed in mid-water. Rays and skates, however, are flattened, and most feed on or just above the seabed. In sharks and rays, the teeth grow in rows that wear away or are lost and

▲ *Southern stingrays live in shallow coastal areas of the Atlantic Ocean and are most abundant off the coast of Florida. This type of stingray can grow to more than 6 feet (1.83 m) wide.*

are replaced continually by new rows at the back. Fertilization of the female's eggs occurs internally, with the male transferring his sperm through a pair of modified fins that form "claspers."

● **Rays and skates** The 500 or so species of rays and skates are known as batoid elasmobranchs (members of the superorder Batoidea). Several obvious features distinguish them from sharks (superorder Selachii). The pectoral fins of batoids are enlarged to form a single broad disk or diamond shape that is attached to the back of the skull. The eyes are positioned on top of the body, rather than on the sides, and batoids all have gill slits on the underside. In sharks, the gill slits are on the sides of the body. These differences can be largely explained by the evolution of sharks as mid-water swimmers and the evolution of rays and skates as bottom-dwellers, although today not all sharks and rays fall into these simple distinctions.

● **Stingrays** These fish and their close relatives belong to the order Myliobatiformes. Some members, such as the eagle rays (family Myliobatidae), have a head clearly distinct from the rest of the body; however, in most stingrays the head is not distinct. Not all members of the group have a stinger, and of those that do, not all are called stingrays. The stinger, when present, is found on the tail and is used in defense and only rarely for attacking prey.

EXTERNAL ANATOMY Rays and skates have enlarged pectoral fins shaped rather like wings, which most species use for propulsion. *See pages 1240–1241.*

SKELETAL AND MUSCULAR SYSTEMS As in sharks, the skeleton is of light cartilage and is supplemented by a tough skin that serves as an exoskeleton for muscle attachment. *See pages 1242–1244.*

NERVOUS SYSTEM Rays and skates have an array of sensory structures, including skin receptors that detect the electrical fields of buried prey. *See pages 1245–1247.*

CIRCULATORY AND RESPIRATORY SYSTEMS Spiracles ensure a flow of clear, oxygen-rich water to the gills, even when the fish is half buried in sand on the sea bottom. *See page 1248.*

DIGESTIVE AND EXCRETORY SYSTEMS Rays and skates have a short gut, containing a spiral structure that slows the passage of food for digestion and increases the surface area for absorption of nutrients. The oil-rich liver helps provide buoyancy. *See pages 1249–1250.*

REPRODUCTIVE SYSTEM Fertilization is internal. Rays bear well-developed live young. Skates lay large, yolky eggs that develop slowly before hatching. *See page 1251.*

External anatomy

In stingrays, as in most other types of rays, the flattened pectoral fins extend forward to the head to form a disk, with sides that look like a pair of wings. The southern stingray swims by sending ripplelike waves along these greatly enlarged fins. The stingray spends much of its time lying on the seabed, waiting in ambush for approaching prey, or cruising just above the seabed and scanning for prey using its skin electroreceptors. The stingray's dark gray, green, or brown coloration on the upper, or dorsal, surface is usually darker than the color of the seabed.

The ray's bottom-living lifestyle has, during millions of years of evolution, resulted in a flattened body form with certain structures located either above or below the disk-shaped body. The eyes are positioned on top of the head, oriented toward the available light streaming from above—which is also the likely direction of attack by predators such as sharks. Two orifices called spiracles, lying just behind the eyes, enable the stingray to take in water for breathing while the mouth is closed. Using the spiracles rather than the mouth for breathing greatly reduces the chance that small particles will enter and damage the gills. Water that exits from the gills is expelled through the five gill slits on the underside of the fish.

▼ **Southern stingray**
The stingray's most striking feature is its body form: a flattened and disk-shaped body with immense pectoral fins and a long tail. Large female stingrays can weigh more than 200 pounds (90 kg).

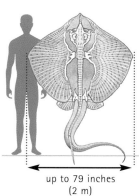

up to 79 inches (2 m)

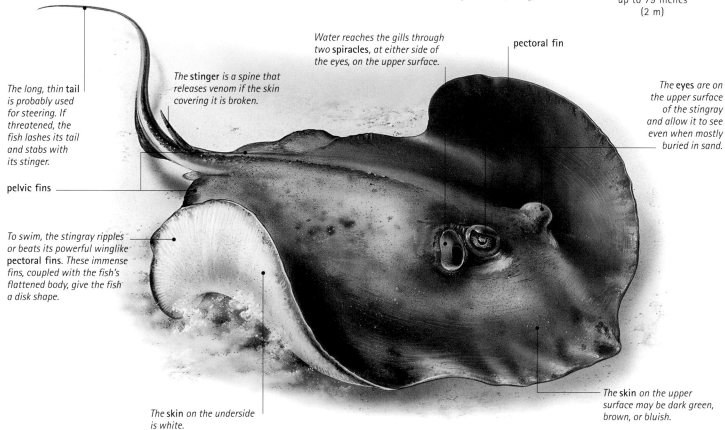

The long, thin **tail** is probably used for steering. If threatened, the fish lashes its tail and stabs with its stinger.

The **stinger** is a spine that releases venom if the skin covering it is broken.

Water reaches the gills through two **spiracles**, *at either side of the eyes, on the upper surface.*

pectoral fin

The **eyes** *are on the upper surface of the stingray and allow it to see even when mostly buried in sand.*

pelvic fins

To swim, the stingray ripples or beats its powerful winglike **pectoral fins**. *These immense fins, coupled with the fish's flattened body, give the fish a disk shape.*

The **skin** *on the underside is white.*

The **skin** *on the upper surface may be dark green, brown, or bluish.*

The sting in the tail

The southern stingray's sting is delivered by a stinger that is a highly modified denticle (toothlike scale) about one-third of the way along the length of the tail. The spine is largely defensive; it is driven forward by the tail arching over the back when the fish is threatened. When the spine is driven into a potential attacker, a thin layer of skin covering the spine ruptures. This action releases venom that has gathered on the spine, supplied by a gland at the base of the spine. The spine is flat and tapered with small serrations along its two edges that anchor the spine in the victim. The venom contains a variety of toxins, including those that produce localized swelling and intense pain, as well as others that affect heart rate. People are sometimes stung when they accidentally step on a stingray in shallow water. The sting is very painful but not normally life-threatening, unless the wound becomes infected with bacteria or the victim is highly allergic to chemicals in the venom. Research into stingray venom for its potential usefulness in medical treatments is ongoing.

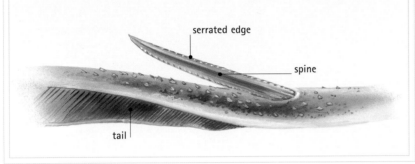

serrated edge

spine

tail

Bony flatfish

Some bottom-living bony fish, such as flounder and turbot, also have a flat form. Their flat body enables them to lie on the seabed, glide just above it, or bury themselves in sediment in which they can remain unseen, waiting for passing prey. Bony flatfish have achieved their flatness in a completely different way from cartilaginous rays. Over millions of years of evolution, the fish have turned on their side. This is evident from the changes that occur when a larval flatfish develops into an adult. The hatchling swims with its body oriented vertically, but during development it gradually tilts onto one side until it swims with its body horizontal. At the same time, one eye gradually moves across to join the other on the same side of the head and the mouth twists into a suitable position for feeding in the horizontal position.

In the southern stingray, any vestige of a dorsal fin is lacking, and the tail, as in many other types of ray, is long and whiplike. On the underside of the fish, the pelvic fins lie on either side of, and just behind, the cloaca. The cloaca is the common opening for the reproductive, digestive, and excretory systems. In males, the inner surfaces of the pelvic fins are modified into tubelike structures—claspers—that are used to transfer sperm to the female during mating. Unlike the fins of most bony fish, which are thin, flexible, semitransparent, and supported by bony rays, those of sharks and rays are fleshy and cannot be folded closely against the body.

Ray skin

In sharks and rays, the sharp, toothlike scales, or denticles, are quite unlike the flattened scales found in most bony fish. In many rays, the denticles are small, and in some species, such as stingrays and electric rays, the skin is very

▼ *The nares (which are used for smell, not respiration) and mouth can be seen on the underside of this spotted eagle ray.*

smooth and the body surface highly flexible. In the southern stingray, some of the denticles develop into short defensive spines that run in two or more rows from front to back behind the eyes. The skin of many species of rays is covered with mucus that smoothes the body surface, helping to streamline the fish.

The mouth of a ray is on the underside, with nostrils, or nares, in front of the mouth. The mouth is armed with rows, or "pavements," of flattened teeth that crush and grind prey, rather than stabbing or chewing them.

Skeletal and muscular systems

CONNECTIONS

COMPARE the pectoral fins of a stingray with those of a *HAMMERHEAD SHARK*. The stingray has huge pectoral fins supported by radials of cartilage. The hammerhead's pectoral fins are relatively much smaller and are supported by struts called ceratotrichia.

Skates and rays are essentially flattened versions of sharks. Sometime, between 250 million and 200 million years ago, some sharks evolved a flattened form better suited for life on or near the sea bottom than for swimming in mid-water. Along with this overall change in body shape, specific adjustments to cope with a diet of bottom-living animals also occurred. The jaws became more flexible, with weaker attachments to the skull than are found in typical shark forms. The teeth became flatter for crushing the shells of invertebrates such as crustaceans and mollusks, rather than pointed

IN FOCUS

Why cartilage?

In common with sharks, skates and rays have a skeleton made of cartilage rather than bone. The ancestors of sharks and rays did have a bony skeleton, but this evolved into cartilage, which is lighter than bone and is also softer and more flexible. The adoption of cartilage rather than bone may be linked to buoyancy. Whereas most bony fish have an air-filled swim bladder that gives the fish buoyancy, sharks and rays do not. Instead, they have a large, oil-rich liver that provides buoyancy, and they have pectoral fins shaped like airfoils, which generate lift. Sharks and rays have thick skin, which is like a corset made of a mesh of tough collagen fibers on the inside and has toothlike denticles on the outside. This thick skin acts like an external skeleton to which the swimming muscles can anchor. This permits the cartilaginous skeleton to be less strong than the skeleton of bony fish, which do not have a tough skin and rely on the bony skeleton for muscle anchorage.

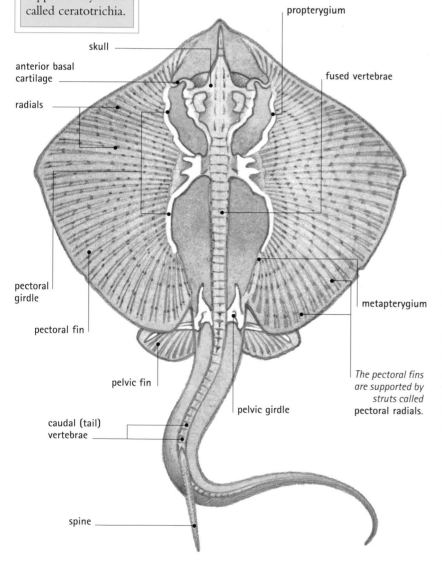

skull

anterior basal cartilage

radials

pectoral girdle

pectoral fin

pelvic fin

caudal (tail) vertebrae

spine

propterygium

fused vertebrae

metapterygium

The pectoral fins are supported by struts called pectoral radials.

pelvic girdle

for grasping fast-swimming fish and squid. At the same time, the openings to and from the gills gradually shifted position so the spiracles could inhale sediment-free water from above the fish while the gill slits exhaled water beneath the fish. Today, these flattened sharklike forms are the skates and rays. Some sharks, such as angel sharks and saw sharks, do have a flattened body and live on or near the seabed, but they differ from skates and rays in the position and arrangement of the gills and the skeletal arrangement in the pectoral fins.

◀ Stingray

The most striking features of the stingray's cartilaginous skeleton are the large pectoral girdle and the fan of radials arising from the girdle. The vertebrae in the main body of the fish are fused to form an inflexible rod.

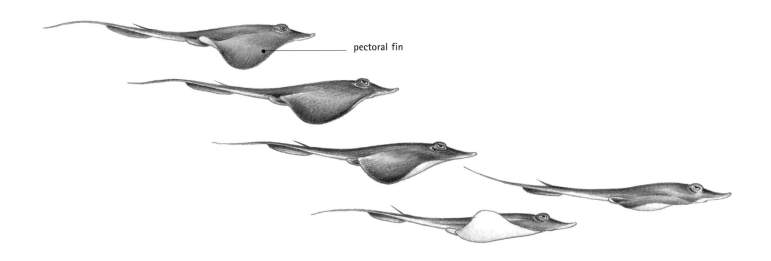

pectoral fin

Compared with those of sharks, the skulls of skates and rays are long from front to back and generally flatter. The jaw mechanism is quite different. Whereas in sharks the jaws swing forward to bite, rays bite by swinging the lower jaw downward. The attachments between skull and jaws are generally weaker in rays, and the jaw construction is lighter.

Stingrays and most skates have pavements of teeth that are slightly pointed for grasping and crushing struggling prey.

Trunk and tail

Whereas most sharks swim by side-to-side movements of the trunk and tail, in rays this swimming action is greatly supplemented, or is entirely replaced, by movements of the enormously enlarged pectoral fins. In rays, the vertebrae in front of the pectoral girdle are joined together as a rigid rod, rather than being flexible as in sharks. Because the tail in skates and rays has a reduced role or no role in swimming, it takes on other functions, such as acting as a defensive weapon, as in stingrays.

In the course of the ray's evolution, the pectoral fins have become enlarged by being supported by a pectoral girdle that is extended from front to back. This articulates with basal elements of the pectoral fin, the front and back elements of which—called the propterygium and the metapterygium, respectively—have become enormously long. The propterygium is also supported by a link of cartilage attached to the skull. The pectoral fins are supported by very long radial cartilages, which are loosely

COMPARATIVE ANATOMY

Rays that swim like sharks

Compared with other rays, guitarfish have relatively small pectoral fins, and their tails, therefore, take on a correspondingly larger role in swimming. They swim by moving the tail and rear portion of the trunk from side to side. To do so, they have blocks of muscle arranged the same way as found in sharks. The dorsal and caudal fins are quite large, and the rear half of the body looks very similar to that of a shark. In comparison with those of other rays, their pectoral fins are stiff and, rather than providing propulsion, they provide stability against rolling.

connected to each other at front and back by fibrous tissue. The ceratotrichials, which are the long supporting struts in the fins of sharks, have been lost entirely in the evolution of rays. The overall effect of having numerous radial elements attached to a broad surface at the base is a broad pectoral fin that is highly flexible in the vertical range, quite unlike the pectoral fins of most sharks.

Swimming actions

Like sharks, the different species of skates and rays live in a wide variety of habitats and often have distinct feeding strategies. Most skates and rays are bottom-dwellers, and they swim just

▲ SWIMMING ACTION
Stingray
The illustration above shows the sequence of movements—at intervals of 0.1 second—of the stingray's pectoral fins. Stingrays move forward by sending ripplelike waves from front to back along the flexible edge of the pectoral fins.

1243

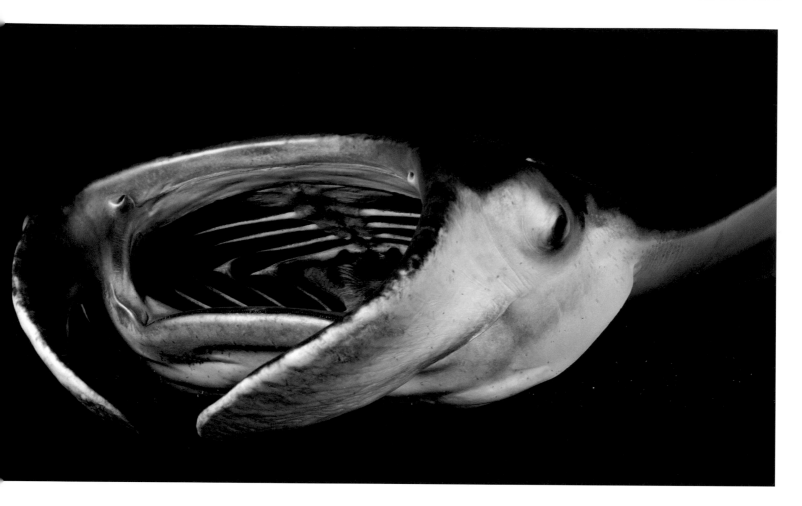

▲ *Manta rays swim in mid-water by flapping their muscular pectoral fins rather like the wings of a bird. When feeding, the manta ray swims with mouth agape, filtering seawater for plankton.*

above the seabed hunting for food or settle on or just below the seabed's surface layer and wait for an unsuspecting victim to pass by. Some rays, such as the southern stingray and the blunt-nose stingray, prefer sandy seabeds, whereas others, such as the blue-spotted stingray and the mangrove whipray, favor coral reefs and other hard bottoms.

More active swimmers

A few species of rays swim more continually in mid-water than typical bottom-dwellers. Cow-nose rays and eagle rays make daily migrations to feeding grounds on the seabed. Manta rays remain almost exclusively in mid-water, where they filter seawater for plankton and small fish. In rays, the greatly extended pectoral fins and flat pelvic fins act like the wings of a glider, giving the fish lift. By bending the rear edge of the pectoral disk downward, the ray can increase lift. By bending one trailing edge up and the other side down, the fish can turn elegantly. The dorsal and caudal fins, if present, are like the tail

fin and rudder of a glider, aiding steering and helping to prevent the fish from rolling.

Skates and stingrays move forward by sending ripplelike waves from front to back along the flexible edge of the disk. These species can reverse the direction of the ripples to stop abruptly and even to move backward. When rays are hovering just above the seabed, or are partially buried in sediment, they can make fine adjustments to parts of the disk and so move in any direction and swivel with greater dexterity than bony flatfish such as sole and turbot.

Mid-water swimmers such as manta rays and eagle rays have powerful muscles attached to the pectoral fins. These contract and relax to flap the outer edge of the disk like the wings of a bird, raising the head and increasing lift with each downstroke. The upstroke is the recovery stroke. Mid-water swimmers can turn by flapping only one side of the disk; and when threatened by sharks, many alternate, flapping one side of the disk and then the other to zigzag away at high speed.

Nervous system

Unlike most sharks, rays generally feed on bottom-living organisms. Rays grab prey with their mouth, which is on the opposite side of the body from the eyes. In hunting, vision is therefore less important for rays than it is for sharks. In common with other cartilaginous fish as well as bony fish, rays can focus on near or far objects by moving their lens backward and forward in the eye. Rays, in general, can see in color as well as in shades of gray. Like sharks, rays have a reflective layer at the back of the eye, called the tapetum lucidum, or "bright carpet." It reflects light back through the eye. In this way, light rays pass through light-detecting sensory cells of the retina not once but twice. This effect almost doubles the sensitivity of the eyes in poor light.

To locate prey at long range, rays, like sharks, rely on a supersensitive sense of smell. They can detect chemicals at concentrations of only one drop in 10 billion. Rays can home in on a scent trail carried many hundreds of yards. They can follow the trail to its source, which could be edible food. As in sharks, the nostrils in rays are blind sacs, technically called nares, that do not connect to the respiratory system. A flap of skin directs the flow of water through the nares in an S-shaped motion, slowing the water's speed for sampling and ensuring that it crosses a very large number of sensory receptors along the way. The olfactory (scent-detecting) cells are packed onto plates of tissue called lamellae, which have a very large surface area. Various substances stimulate different batches of cells by varying amounts, triggering streams of nerve impulses that are transmitted to the brain, where they are interpreted. In this

CONNECTIONS

COMPARE the stingray's electrosensory ability with that of a **PLATYPUS**. Both animals have electroreceptors to detect potential prey and predators. The stingray's electroreceptors are all over its body, whereas those of the platypus are located in its bill.

▼ Rays, such as this blue-spotted ray, have color vision for bright light and monochromatic vision for dim light. In hunting, however, the sense of smell is more important.

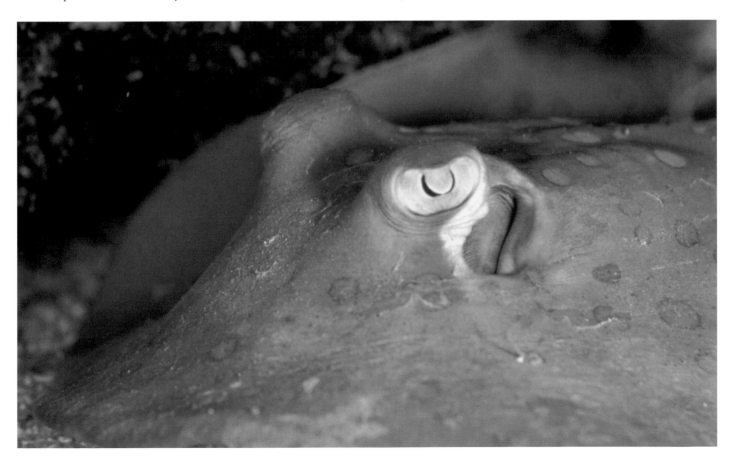

Flat but highly charged

Electric rays, also called torpedo rays, have highly specialized hunting strategies. Some species lie in wait on the seabed, covered in a loose layer of sediment. When a fish comes close, the ray begins to deliver powerful electric pulses of 50 to 220 volts and up to 8 amps, immobilizing the victim. Simultaneously, the ray rises from the sea bottom, and this action "sucks" the victim under the pectoral disk. The ray continues to deliver the shocks for as long as a minute, in which time it maneuvers the stunned victim to the mouth. Predators and humans also receive shocks from electric rays. The shocks are powerful enough to knock people off their feet.

The electric ray's electrical discharges are generated by kidney-shaped patches of muscle tissue that were originally associated with the gills. These electric organs, weighing about 15 percent of the mass of the fish, lie in the pectoral fins and contain 200 to 1,000 vertical columns of cells, called electroplaques. The cells in each electroplaque are arranged end to end like batteries connected in series. The change in electrical potential across the membranes of each cell is generated by an inrush of ions such as sodium (Na^+), calcium (Ca^{2+}), and chloride (Cl^-). These ions are pumped across cell membranes to reset the mechanism. One mystery is how electric rays can "fire" their cells simultaneously with minimal delay between cells at the top of the electroplaque and those at the bottom, so a powerful, short-lived shock is generated.

▶ ELECTRIC ORGAN
Electric ray
The electric organs are two kidney-shape patches of muscle tissue in the pectoral fins. Electric cells in the tissue discharge simultaneously to produce a strong current that can stun prey in the surrounding water.

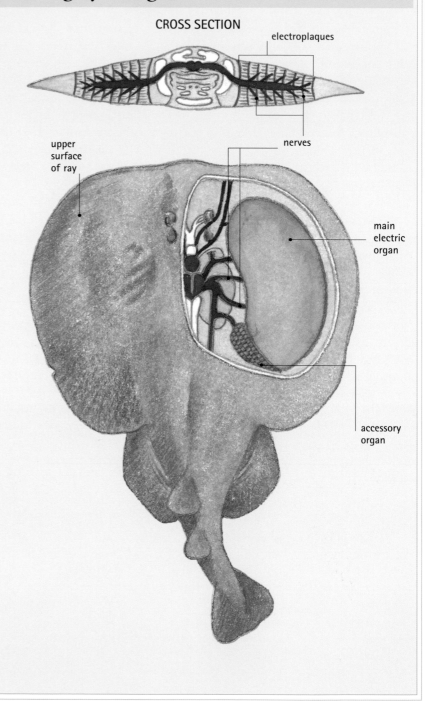

CROSS SECTION

electroplaques

nerves

upper
surface
of ray

main
electric
organ

accessory
organ

way, the ray can probably distinguish hundreds of different scents to identify food or recognize a mate. Like other cartilaginous fish, and bony fish, too, rays taste food items before swallowing. To taste, rays use groups of sensory cells called taste buds that are scattered on the lining of the mouth and tongue.

Touch at a distance
Like sharks, rays can detect objects that touch the skin or mouth directly as well as from pressure waves or vibrations that travel through the water. Effectively they have a "touch-at-a-distance" sense. Various sensory receptors in the dermis of the skin—the layer

Large brain

Like sharks, rays and skates have a large brain relative to the size of their body. Size for size, most rays have a slightly larger brain than sharks have, and rays' brains are equivalent in size to the brains of marsupials and birds. The large size can be partly accounted for by the remarkable array of sensory receptors sharks and rays have, such as those that detect electric fields. These senses require a large brain to interpret the streams of data they generate. The forebrain—the part of the brain that processes information from olfactory, electrosensory, and lateral line systems—is unusually large in stingrays and manta rays.

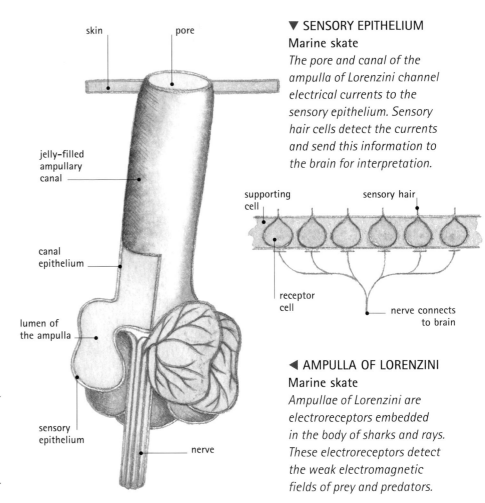

▼ SENSORY EPITHELIUM
Marine skate
The pore and canal of the ampulla of Lorenzini channel electrical currents to the sensory epithelium. Sensory hair cells detect the currents and send this information to the brain for interpretation.

◄ AMPULLA OF LORENZINI
Marine skate
Ampullae of Lorenzini are electroreceptors embedded in the body of sharks and rays. These electroreceptors detect the weak electromagnetic fields of prey and predators.

under the outer layer, or epidermis—respond to direct pressure and other kinds of deformation of the skin, as when the body bends during swimming.

The system that detects vibrations and pressure changes in the surrounding water is called the lateral line system. It consists of a network of small canals that open onto the skin surface through tiny pores. The canals contain a gel-like fluid that moves in response to the arrival of pressure waves. Sensory organs within the canal system, called neuromasts, detect this movement and relay the information to the brain. This system gives the fish a picture of the objects in its vicinity that are disturbing the water. Rays and sharks can recognize disturbances from several hundred feet away.

In sharks and bony fish, the lateral line system is visible as a line running along the middle of each flank. In rays, the branches of the system are scattered over the upper and lower surfaces of the pectoral fins, as well as on the head. In stingrays, part of the system is concentrated in the tail to give early warning of shark attacks from the rear.

The electrical sense
Sharks and rays can detect electric fields, and this supersense is one of their most useful and intriguing abilities. Scattered over the body of skates and rays are numerous pores, each

leading by way of a gel-filled canal to a chamber, the ampulla of Lorenzini, which resembles a miniature bunch of grapes. Cells lining the walls of the ampulla can detect the electrical fields that all living organisms produce. This electromagnetic field is created by the small voltage differences across the surface of their cells. These voltage differences are largest and most widespread in muscle tissue, and so even a stationary fish, crustacean, or worm can be detected from the muscles that contract inside it during the process of breathing.

Rays are able to detect the electrical signature of camouflaged and buried prey at distances of up to 12 inches (30 cm). During the breeding season, male round stingrays (genus *Urolophus*) use their electrical sense to detect the recognizable electrical field of females. Moreover, some female stingrays lie buried in the sediment of the seabed, waiting to be electrically detected by a suitable mate.

Circulatory and respiratory systems

COMPARE the circulatory system of a fish such as the stingray with that of a mammal, such as the *ZEBRA*. Fish have a single circulation, in which blood passes only once through the heart in each full circuit of the body. Mammals have a double circulation, in which the blood passes through the heart twice in a full circuit.

Rays have a muscular heart that pumps blood around the body in a single circuit, delivering dissolved foods and oxygen to body tissues. From the heart, the blood travels to the gills, where it picks up oxygen and gets rid of carbon dioxide, before circulating to other body tissues and then returning to the heart at lower pressure.

Like sharks, rays use gills to get the oxygen they need. Carbon dioxide, which is a waste product of respiration, is exchanged for oxygen. The exits to the gills are located on the lower surface of the ray.

Most sharks take in water through the mouth in order to provide the flow across the gills for gas exchange. However, rays and some bottom-living sharks use their spiracles as well. These openings lie behind the eyes and take in water from above the fish. When a fish is half-buried in sand on the seabed, using the mouth would draw in oxygen-poor water laden with sand particles. Drawing in water through the spiracles ensures a clear, oxygen-rich supply.

▼ BREATHING
Most fish "breathe" water that they take in through the mouth and expel through the gill slits. Rays take in water through their spiracles (the holes behind each eye) and expel the wastewater through their gill slits on the underside. The flow of water prevents sand and mud from clogging the gills.

Fish hearts

The heart of a shark or ray is very similar to that of a bony fish. Both have two proper heart chambers—the atrium and ventricle—and two associated structures. The pumping action of the heart is carried out mainly by the atrium, which raises blood pressure slightly, and then the ventricle, which raises the pressure greatly. One of the associated structures—the sinus venosus—in both cartilaginous fish and bony fish develops from modified tissue of the major veins leading to the heart. The sinus venosus, by bulging with blood and then elastically recoiling, helps ensure a continual flow of blood to the atrium. In bony fish, the bulbus arteriosus is a thickened part of the aorta, the main artery leading from the heart. It smoothes the pumping action from the ventricle, so the blood is pumped almost continuously around the body without inefficient stops and starts. Sharks and rays have a similar structure—the conus arteriosis—but it develops from heart muscle and is fibrous and equipped with valves to prevent blood from flowing back into the heart.

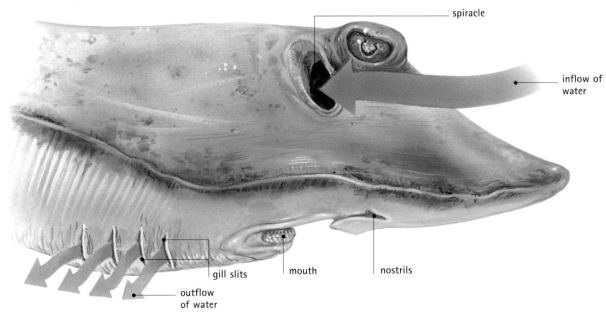

spiracle

inflow of water

gill slits

mouth

nostrils

outflow of water

Digestive and excretory systems

Rays may have to wait several days between one meal and the next. To accommodate large prey items, rays have stretchable stomachs. Digestion begins in the stomach and continues in the intestine. Both secrete digestive enzymes from their linings. The intestine also receives bile from the liver and digestive enzymes from the pancreas. Sharks, rays, and skates have an unusually short intestine compared with that of bony fish, and they compensate for this by incorporating a spiral valve. This corkscrew-shape structure greatly slows the passage of digested food through the gut, in addition to increasing the surface area across which the products of digestion are absorbed.

The intestine's blood supply absorbs products of digestion, which soon travel to the liver, where excess glucose is temporarily stored as glycogen along with fat-soluble vitamins such as vitamins A and D. The remaining products of digestion are distributed throughout the body, where they are incorporated into new cell parts

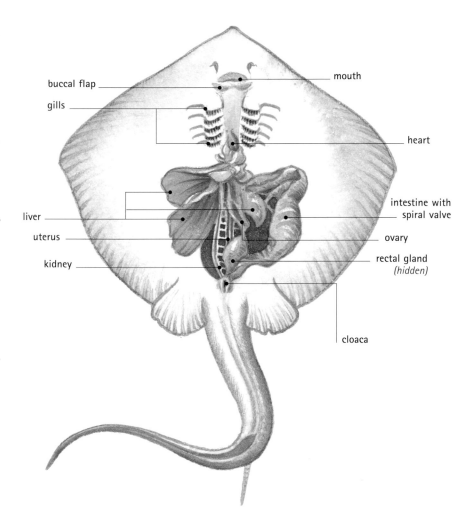

Labels on diagram: buccal flap, gills, liver, uterus, kidney, mouth, heart, intestine with spiral valve, ovary, rectal gland (hidden), cloaca

▲ Stingray
Compared with bony fish, rays have a short intestine. However, the spiral valve in the intestine increases the surface area for absorption and slows the transit of food through the gut, allowing maximum absorption.

PREDATOR AND PREY

The devil ray

The manta ray is also called the devil ray—so named because of its "horned" head. It is the largest ray, reaching an astonishing 30 feet (9 m) across and weighing up to 3,100 pounds (1,400 kg). Manta rays feed on small, schooling fish and planktonic crustaceans that they filter from the water. The paddlelike extensions at the front of the manta ray's head, called cephalic lobes, are modified regions of the pectoral fins. The lobes help guide water into the manta ray's open mouth. Animals are extracted from the incoming water as it flows through the gills. This extraction is done by spongy tissue that bridges the gap between one gill arch and the next. When manta rays find a food-rich patch of water, they will sometimes swim in slow somersaults to stay in the hot spot and harvest as many prey as possible.

IN FOCUS

A buoyant liver

Because sharks and rays do not have a swim bladder—unlike bony fish—they need other methods to make their body more buoyant; otherwise, they would expend considerable amounts of energy swimming upward to avoid sinking. Buoyancy is achieved by having a large liver packed with oil. The oil weighs about 90 percent of the equivalent volume of salty water, and so lightens the load inside the fish. But sharks and rays also use another mechanism: trimethylamine oxide (TMO) and urea dissolved in the blood also contribute to buoyancy.

▶ *The yellow stingray, like most rays, has a mouth on the underside of its body, behind the nostrils. This ray is a bottom-feeder; it grubs around on the seabed for worms, crustaceans, and mollusks buried in the sediment.*

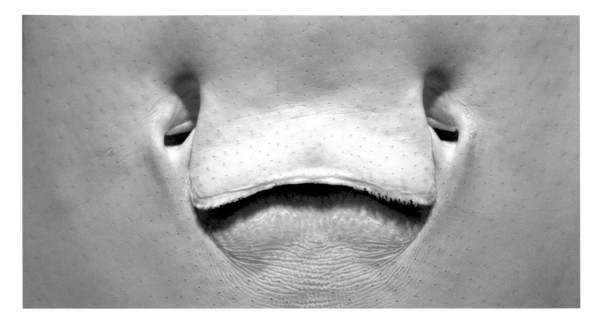

▼ TEETH
Eagle ray and thornback ray
The eagle ray's pavements of broad, strong teeth are suited to grinding the shells of shellfish. The small, pointed teeth of the thornback ray are suited to crushing soft-bodied worms and small shelled animals.

or provide fuel for respiration. Undigested matter remaining in the gut exits through the anus.

Sharks, rays, and skates maintain a lower concentration of salts than is found in seawater. Two kidneys help remove excess salts that are taken in with the food. Sharks and rays have an additional structure, the rectal gland, that gets rid of salts through the anus. The rectal gland functions like a third kidney. To slow the entry of salts into the fish, and the exit of water from the fish by osmosis, sharks and rays pack their blood with two substances—urea and trimethylamine oxide (TMO). These bring the level of dissolved substances in the blood to levels similar to those found in seawater.

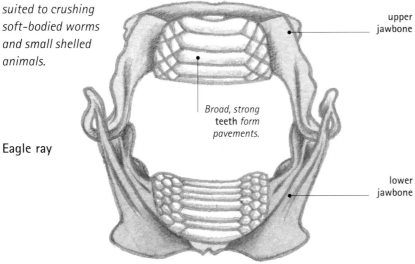

Eagle ray

upper jawbone

Broad, strong teeth form pavements.

lower jawbone

Thornback ray

upper jawbone

small, pointed teeth

lower jawbone

COMPARATIVE ANATOMY

Crushing teeth and jaws

Teeth provide clues to the feeding strategies of rays. The teeth of the thornback rays and diamond stingrays are small and pointed for dealing with soft-bodied animals, such as sandworms, and relatively small hard-shelled animals, such as shrimp and small crabs. The teeth of eagle rays and cow-nose rays, on the other hand, are much broader and stronger and are suited to crushing the stouter shells of mollusks such as oysters, clams, and sea snails.

Recent research shows that the teeth of the cow-nose ray are strengthened by deposits of calcium, giving the teeth bonelike strength. The cow-nose ray positions its prey in the middle of its jaws and contracts the jaw muscles on one side while the other side is braced, acting as a pivot. The mechanism is like that of a nutcracker. By using this type of lever system, the ray can exert enough force to crack open thick shells.

Reproductive system

The mode of reproduction is an important distinction between skates and rays. Skates lay large yolky eggs, a reproductive strategy called oviparity. Rays bear live young, and the young develop inside the mother but are not attached to her by a placenta, a strategy called ovoviviparity.

Male rays and skates, like sharks, fertilize females internally by using their claspers—the modified inner edges of the pelvic fins. During mating, the male aligns the claspers to form a tubelike structure that is inserted into the female's cloaca. Through this, the male's sperm is transferred to the female.

Courtship in the southern stingray proceeds with one or more males closely following a female for many minutes. Males are much smaller than females: typically about half the females' width. Prior to mating, the male grasps the female's pectoral disk in his mouth and then swings beneath the female so that their ventral surfaces are facing each other. He then inserts his claspers into the female's cloaca. Copulation lasts for about 30 seconds, after which the male lets go. It is quite common for another male to mate with the female soon after.

Stringlike trophonemata

In the diamond stingrays, butterfly rays, and eagle rays, the developing embryos initially get their nutrients from yolk sacs. Later, however, the pregnant mother nourishes the embryos with a protein-rich "uterine milk" that she delivers through stringlike extensions of her uterus, called trophonemata. These are inserted through the spiracles of the embryos. The gestation period lasts five months, after which three to five stingray young are born. They have soft denticles and a flexible stinger that cannot harm the mother during birth.

Skates do not give birth to live young but lay rectangular leathery egg cases. The egg cases have long tendrils at each corner and contain an embryo. The female anchors the egg cases to the seabed. The surface of the egg case is usually sticky so that sediment and other debris stick to it and weigh it down. Embryos take several

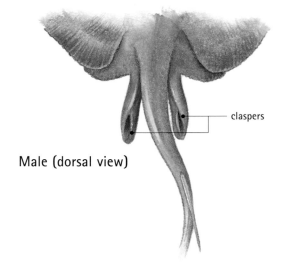

Male (dorsal view)

claspers

CONNECTIONS

COMPARE the trophonemata of a stingray with the placenta of a **LION**. The trophomemata supplies the embryos with a protein-rich "milk," whereas the placenta allows oxygen and nutrients to pass from the mother's blood to the fetus's blood.

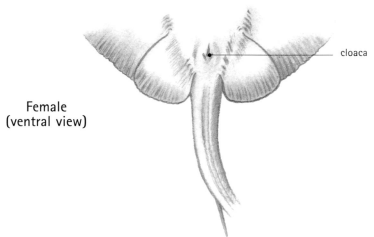

Female (ventral view)

cloaca

▲ EXTERNAL REPRODUCTIVE STRUCTURES
Stingray
During mating, the smaller male aligns his claspers into a tubelike structure, which he inserts into the female's cloaca to transfer sperm. In rays, embryos develop in the mother and are born about five months after fertilization.

months to develop. The young inside grow by consuming the enclosed yolk supply. The egg cases are tough and have an unpleasant taste and texture, thus deterring predators.

TREVOR DAY

FURTHER READING AND RESEARCH
Hamlett, W. C. (ed.). 1999. *Sharks, Skates, and Rays: The Biology of Elasmobranch Fishes.* Johns Hopkins University Press: Baltimore, MD.
Moyle, P. B., and J. J. Cech. 2000. *Fishes: An Introduction to Ichthyology* (4th ed.). Prentice Hall: Upper Saddle River, NJ.
Paxton, J. R., and W. N. Eschmeyer (eds.). 1998. *Encyclopedia of Fishes* (2nd ed.). Academic: San Diego, CA.

Tarantula

PHYLUM: Arthropoda CLASS: Arachnida ORDER: Araneae
SUBORDER: Mygalomorphae FAMILY: Theraphosidae

Tarantulas are among the largest spiders on Earth, with a leg span up to 11 inches (28 cm). These eight-legged invertebrates prey on insects and small vertebrates, and many species live in silk-lined burrows in the ground.

Anatomy and taxonomy

Scientists group all organisms into taxonomic groups based largely on anatomical features. Spiders are divided into three suborders, and tarantulas belong to the suborder Mygalomorphae, which contains 11 families and many genera and species.

● **Animals** Animals are multicellular organisms with cells that are generally organized into tissues and organs. Both animal and plant cells contain membrane-bound structures called organelles (mini-organs). Unlike plants, however, which make their own food, animals must eat to get energy. Almost all animals have guts that enable digestion, and most animals have a nervous system that controls how they interact with their surroundings.

● **Arthropods** These form the largest and most successful phylum of organisms, yet despite their vast diversity the general body plan of arthropods is relatively constant. They have segmented bodies, although segments are often fused to form units such as the head and abdomen. Arthropods have a tough outer exoskeleton. It protects the internal organs and serves as an attachment point for muscles. To grow, an arthropod must shed, or molt, its exoskeleton. Arthropods have pairs of jointed legs, mouthparts, and antennae. Internally, all arthropods have a ventral nerve cord, which runs near the underside of the animal. The rear part of a dorsal vessel pumps a bloodlike liquid called hemolymph around the body cavity.

● **Chelicerates** Organisms in the subphylum Chelicerata have a body with two main divisions: the prosoma and the opisthosoma. The prosoma bears six pairs of appendages.

▼ *Scientists still debate the relationship of different spider families. There are around 15 families of mygalomorph spiders, including the trap-door spiders, funnel-web spiders, purse-web spiders, and tarantulas. There are around 70 families of araneomorph spiders and just one family of giant trap-door spiders.*

Animals
KINGDOM Animalia

Arthropods
PHYLUM Arthropoda

Crustaceans
SUBPHYLUM Crustacea

Chelicerates
SUBPHYLUM Chelicerata

Uniramians
SUBPHYLUM Uniramia

Horseshoe crabs
CLASS Merostomata

Arachnids
CLASS Arachnida

Sea spiders
CLASS Pycnogonida

Mites and ticks
KINGDOM Acarina

True spiders
ORDER Araneae

Scorpions
ORDER Scorpiones

Araneomorphs
SUBORDER Araneomorphae

Mygalomorphs
SUBORDER Mygalomorphae

Giant trapdoor spiders
SUBORDER Mesothelae

Tarantulas
FAMILY Theraphosidae

In a typical chelicerate the first of these are claws, or chelicerae. The second pair of appendages, or pedipalps, are also modified in various ways. In scorpions, for instance, the pedipalps form the large pair of claws, whereas the true chelicerae are inconspicuous structures near the mouth.

● **Arachnids** Spiders, scorpions, mites, and ticks are chelicerates belonging to the class Arachnida. Arachnids have four pairs of segmented legs.

● **Spiders** There are at least 34,000 species of spiders, classified in the order Araneae and organized into three suborders and about 86 families. All spiders are carnivores. Spiders have chelicerae—fangs with poison glands—while the pedipalps of the males are modified for mating.

Spiders in the suborder Mesothelae have a clearly segmented opisthosoma. Biologists believe that all other spiders evolved from members of this suborder, which is now reduced to just a single family. They live in a burrow, the entrance to which is covered with a trapdoor. More than 90 percent of spiders belong to the suborder Araneomorphae. These spiders are sometimes called true spiders. Their chelicerae are attached to the prosoma below the head and are used in a side-to-side action to increase their biting span. Breathing is through book lungs and tracheae.

Tarantulas were named for a wolf spider, *Lycosa tarantula*, whose bite people mistakenly believed spread the "disease" tarantism in southern Europe. The illness was first recorded

▲ *The goliath tarantula is the world's largest spider. An adult female may be have a leg span of 10 inches (25 cm) or more and a body almost 5 inches (12 cm) long.*

in medical journals in the 14th century, and it was believed that the only cure was to dance to certain music: the tarantella. *Lycosa tarantula* is a member of the family Araneomorphae, but when people refer to tarantulas they usually have in mind the large spiders in the third major group, the suborder Mygalomorphae.

● **Mygalomorphs** Spiders in the suborder Mygalomorphae are distinguished by their chelicerae, which are attached to the front of the head and strike forward and down. Breathing is through book lungs only. There are 15 families of mygalomorphs, including trap-door spiders, funnel-web spiders, purse-web spiders, and tarantulas.

● **Tarantulas** There are about 850 species of tarantulas in the family Theraphosidae. Most live in North and South America, but there are others in Africa, Asia, and Australia.

External anatomy

CONNECTIONS

COMPARE the eight jointed walking legs of a tarantula with the six legs of an insect such as an **ANT** or a **HONEYBEE**.

COMPARE the two-section body of a tarantula with that of an **ANT**. A tarantula has a prosoma and an abdomen, whereas an ant has a head, thorax, and abdomen.

All spiders have an exoskeleton, or cuticle, and their body has two main sections: a prosoma, or cephalothorax, at the front; and an abdomen, or opisthosoma, behind. The two sections are connected by a narrow stalk called a pedicel. All spiders have four pairs of legs and two pairs of other long appendages (a pair of chelicerae and a pair of pedipalps), which are attached to the prosoma.

Spiders vary greatly in size, but most species are relatively small. One Bornean species has a body just 0.01 inch (0.37 mm) long when full-grown. However, tarantulas are relatively large, and the biggest—the goliath tarantula—can cover a dinner plate when its legs are extended. The combined length of its prosoma and abdomen is 5 inches (12.5 cm), and its legs may span 1 foot (30.5 cm).

The prosoma is covered by two rigid plates: the carapace on the upper side and the sternum underneath. The carapace has an indentation along its midline. Together, the carapace and sternum form a relatively rigid exoskeleton. In tarantulas and most other spiders, the abdomen is unsegmented and relatively soft. There are spinnerets at its rear end, where a spider's silk threads emerge from the body. Four slits penetrate the underside of the abdomen, and it is through those that the spider draws in air.

Six pairs of appendages attach to the prosoma: there are four pairs of jointed walking legs; one pair of biting chelicerae, which in tarantulas bear strong fangs; and one pair of pedipalps, which in mature males operate as mating organs.

The first pair of appendages on the prosoma are called chelicerae. The chelicerae are in two sections: a stout basal section and a movable fang. In many spiders, the inner edge of the fang is serrated, allowing the spider to manipulate silk threads. Tarantulas and most other spiders use their chelicerae to inject venom into prey. Usually, each fang rests in a groove in the base, but when the spider bites prey the fangs move out a little like the blade of a penknife. Poison then passes along a duct in the fang and is injected from a hole at its sharp end. Other spiders use the appendages to carry egg cocoons or prey, to dig burrows, or to grasp their partner during mating.

Pedipalps and legs

The second pair of appendages are called pedipalps. Each has six segments: nearest the body are the short coxa and trochanter; then come a long femur, a kneelike patella, a tibia, and a tarsus. The pedipalps are sensitive feelers, which can "taste" food and manipulate prey. Males also use the appendages to transfer sperm to the female, a behavior unique to spiders. The two coxae have evolved to some degree into mouthparts called maxillae.

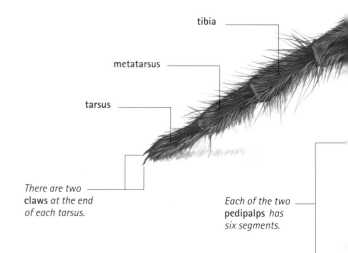

tibia

metatarsus

tarsus

There are two **claws** *at the end of each tarsus.*

Each of the two **pedipalps** *has six segments.*

▶ **Goliath tarantula**
The basic body structure of a tarantula is the same as for other spiders. There are two body sections: a prosoma and an opisthosoma. Six pairs of jointed appendages attach to the prosoma: four pairs of legs, one pair of pedipalps, and one pair of chelicerae. Touch-sensitive hairs cover the body.

The other appendages on the prosoma are the four pairs of legs. They fan out from the pliable connection between the carapace and the sternum of the prosoma. Each leg has seven segments, which are the same as the segments of the pedipalps with one addition: there is a metatarsus between the tibia and the tarsus. The legs are able to articulate at the junctions between the segments, enabling the spider to raise and lower them, and thus walk. Tarantulas have two claws at the end of the tarsus, and some other spiders have three claws. Between the claws are are thick tufts of hairs called scopulae. Some tarantulas have a more extensive covering of scopulae extending along the ventral surface (underside) of the tarsus and metatarsus. The tip of each hair divides into thousands of tiny filaments, an arrangement

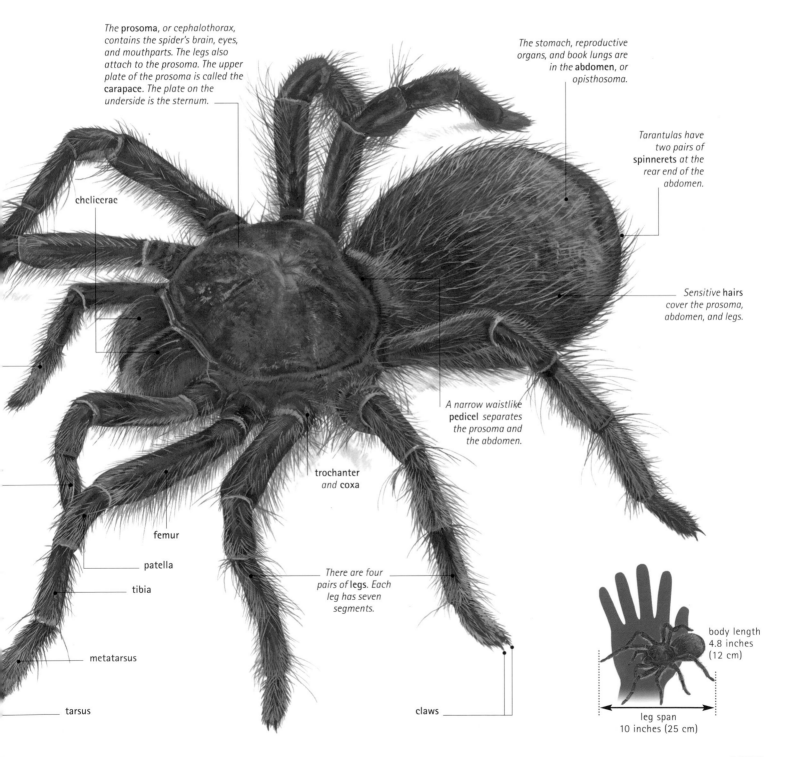

*The **prosoma**, or cephalothorax, contains the spider's brain, eyes, and mouthparts. The legs also attach to the prosoma. The upper plate of the prosoma is called the **carapace**. The plate on the underside is the sternum.*

*The stomach, reproductive organs, and book lungs are in the **abdomen**, or opisthosoma.*

*Tarantulas have two pairs of **spinnerets** at the rear end of the abdomen.*

chelicerae

*Sensitive **hairs** cover the prosoma, abdomen, and legs.*

*A narrow waistlike **pedicel** separates the prosoma and the abdomen.*

trochanter *and* coxa

femur

patella

tibia

*There are four pairs of **legs**. Each leg has seven segments.*

metatarsus

tarsus

claws

body length 4.8 inches (12 cm)

leg span 10 inches (25 cm)

1255

▲ *The sharp, curved fangs are clearly visible on the chelicerae of this tarantula. Poison from poison glands passes along tubes in the fangs and is injected into prey.*

that enables the spider to grip smooth surfaces. The pressure of fluid in the spider's body splays the tufts, increasing the area of filaments in contact with the ground. A crab spider has only about 30 hairs on each foot, yet each foot has 160,000 contact points, providing very strong physical adhesion. Scientists discovered that one spider on a vertical sheet of glass could hold 10 times its weight. This ability helps a tarantula hunt on surfaces that would not be accessible to many other animals.

At the rear of a spider's abdomen are two, three, or four pairs of spinnerets, which "spin" the spider's silk. Most spiders have three pairs, but tarantulas have only two pairs. Although short, the spinnerets are mobile: those of a tarantula move in step with the legs. The spinnerets can be moved independently, and they can also be coordinated; this ability allows the spider to spin complex structures with its silk. Tarantulas use their spinnerets to construct sacs for their eggs. Many other kinds of spiders, but not tarantulas, spin webs to ensnare prey.

CLOSE-UP

Defensive hairs

Most tarantulas are very hairy spiders. The prosoma, abdomen, and legs are covered with sensitive, innervated (nerve-containing) hairs. They provide the spider with very detailed information about its environment. Many of the hairs on a tarantula's body serve another purpose: defense. When faced with a predator, a tarantula can rub small urticating hairs off the abdomen with its hind legs. The urticating hairs have hundreds of tiny hooks that can cause itching. This action is very effective: the hairs can work themselves 0.08 inch (2 mm) into human skin. Goliath tarantulas also have hairs with either hooks or filaments on the longest segments of the front two pairs of legs. If the tarantula is threatened, it repeatedly attaches and detaches the hooks and filaments, producing a hissing sound that may warn off the aggressor.

Internal anatomy

A tarantula's prosoma contains the brain, eyes, legs, and mouthparts, and its prime functions are organizing the central nervous system (CNS), eating, and locomotion.

The anterior section of the digestive tract runs from the mouth through the esophagus to the stomach (which in spiders is called the sucking stomach) and the intestine, which runs into the opisthosoma. A branch of the intestine called the midgut diverticulum runs forward into the prosoma. The tarantula's two-part "brain" is made up of two ganglia (masses of nervous tissue): a supraesophageal ganglion and a subesophageal ganglion. The simple brain connects via optic nerves to the eyes and via the abdominal nerve to the opisthosoma.

A poison gland exits the prosoma in chelicerae adjacent to the mouth. The prosoma contains some important muscles, notably the pharynx muscles, which open and close the mouth. Muscles attach to the cuticular ridge that runs laterally along the midline inside the carapace. A hemolymph (blood) vessel, the intestine, and the abdominal nerve run through the pedicel to connect the prosoma and the opisthosoma. The opisthosoma contains further organs for digestion, circulation, respiration, excretion, silk production, and reproduction. The intestine branches out into lobed structures called diverticula and finally to a stercoral pocket. It is there that digestion occurs. Malpighian tubules carry waste products from the tarantula's hemocoel into a short hindgut and then to the anus for expulsion.

Between the gut diverticula there are other important structures, including a heart, which pumps hemolymph around the body. The book lungs draw air through slits in the ventral side of the opisthosoma and allow oxygen to pass into the hemolymph in the heart. Ovaries in females, or testes in males, exit the opisthosoma at a genital opening on the underside; and the spinning glands produce silk, which leaves the body at the spinnerets. The opisthosoma also contains muscles.

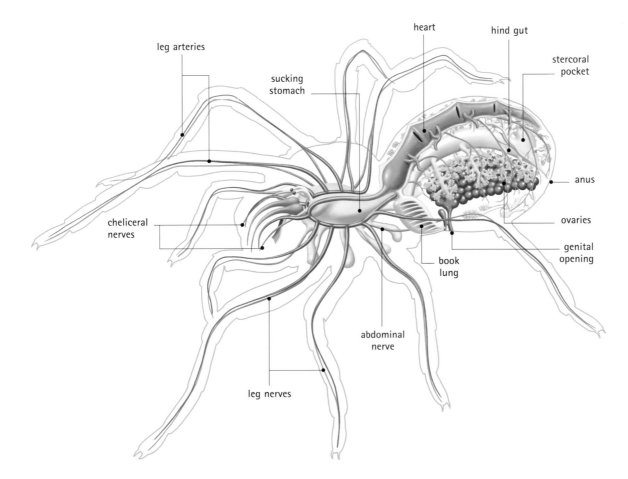

leg arteries

sucking stomach

heart

hind gut

stercoral pocket

cheliceral nerves

anus

ovaries

genital opening

book lung

abdominal nerve

leg nerves

◀ The prosoma (fused head and thorax) contains the brain, eyes, muscles, and part of the digestive tract and is attached to the appendages—the legs and mouthparts. The opisthosoma contains the rest of the digestive tract, book lungs, heart, reproductive organs, and spinning glands.

Skeletal and muscular systems

Like all arthropods and all other spiders, tarantulas have a tough outer layer, or exoskeleton. A spider's exoskeleton is called its cuticle. It is a laminated composite structure that is both strong and elastic and is made of chitin fibers. Chitin is a tough protein. On the prosoma, the cuticle has four layers. The very thin outermost layer is the epicuticle. Since it is on the outside of the spider's body, the epicuticle suffers wear and tear and has to be renewed regularly by material secreted along pore canals from the epithelium, which lies beneath the cuticle. Beneath the epicuticle, in order, are the much thicker exocuticle, the mesocuticle, and the endocuticle. The structure of the cuticle of the opisthosoma is different: most of its thickness is mesocuticle, and the exocuticle is absent entirely. Thus the cuticle of the opisthosoma is not as tough as that of the prosoma.

Parts of the exoskeleton extend into the tarantula's body. These extensions, or apodemes, serve as points of attachment for muscles. For example, the tergal apodeme projects into the prosoma from below the midline groove of the carapace. The muscles of the sucking stomach

Muscles of the leg

Most joints are dicondylous—that is, they move only in one plane. However, the joint of the coxa and trochanter can move forward and backward. Most joints have several muscles: flexors bend the joint and extensors stretch it. The musculature of a tarantula's leg is very complex. For example, species in the genus *Eurypelma* have 30 muscles in each leg. The muscle is striated (skeletal) muscle.

▼ MUSCLES OF SUCKING STOMACH
When the muscles that connect the sucking stomach with the carapace and endosternite contract, the stomach expands. Expansion creates a vacuum that causes the stomach to act as a suction pump.

attach to this projection. Spiders also have an internal skeleton (endoskeleton), the elements of which are called endosterna. The largest section of the endoskeleton is the endosternite; this bowl-like structure with projections is situated in the middle of the prosoma. Limb and stomach muscles attach to the projections of the endosternite. There are smaller endosterna in the tarantula's opisthosoma.

Muscles

A tarantula is dependent on muscles for locomotion, respiration, injecting poison, feeding, and other functions. The cells of the muscles in the legs contain relatively few mitochondria, and since mitochondria provide the energy for the muscles, spiders tire quickly. After a period of activity—for example, if a tarantula runs in pursuit of prey—the spider becomes exhausted and will need to rest.

The muscles of the pharynx and stomach are important in feeding. Food is sucked in through the mouth by the action of those muscles. Several strong bands of muscles connect the upper stomach wall with the dorsal apodeme and the lateral stomach wall with the endosternite. When these muscles contract, the volume of the stomach increases, and the spider sucks. Other muscles work against the muscles that expand the stomach, causing it to contract. By alternating their contractions, the two sets of muscles transform the stomach into an efficient sucking pump. The tarantula is thus able to take in food quickly.

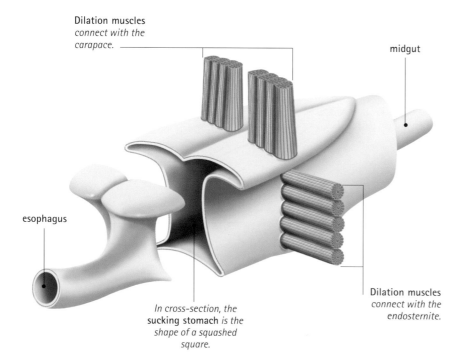

Dilation muscles *connect with the carapace.*

midgut

esophagus

In cross-section, the **sucking stomach** *is the shape of a squashed square.*

Dilation muscles *connect with the endosternite.*

Nervous system

Most arthropods have a central nervous system (CNS) that consists of a simple brain and a ventral nerve cord. The nerve cord is made up of connected ganglia, or masses of nerve cells. Spiders, however, do not conform to this general plan. Their CNS consists of two ganglia in the prosoma: a relatively large supraesophageal ganglion and a smaller subesophageal ganglion, separated horizontally by the esophagus. The center of a spider's CNS is the supraesophageal ganglion, the spider's "brain." This ganglion has nerves connecting to the eyes and the nerve centers controlling the jaws. Nerve fibers run on either side of the esophagus to the subesophageal ganglion, which has neurons supplying the palps, the legs, and various structures in the opisthosoma.

Nerves connect the CNS with the legs and other appendages, sensors, and organs of the opisthosoma. The subesophageal ganglion consists of the fused leg ganglia. The ganglia are connected by many interneurons. Cheliceral ganglia provide the nerves for the muscles of the chelicerae, pharynx, and poison glands. The CNS makes up just 0.1 percent of a tarantula's weight, but 5 percent of a jumping spider's weight. The difference is largely a product of a jumping spider's much larger eyes.

All spiders require a constant flow of information about their external environment for feeding, avoiding predators, and mating, and tarantulas are no exception. They also need an awareness of what is happening within their own body. Tarantulas, just like other spiders, are equipped with a range of receptors to gather this information. Three or four pairs of eyes on the prosoma provide visual information, and chemoreceptors allow them to "taste" and "smell" their environment. Mechanoreceptors allow them to feel what is going on around them, and thermoreceptors give information about temperature. Some other spiders have fewer eyes, and some cave-dwellers have none.

For tarantulas, mechanoreceptors are the most important sensors. They pick up touch, vibrations, and air currents, so the spider is

COMPARE the single ventral nerve cord of a tarantula with the double ventral nerve cord of a *GIANT CENTIPEDE*.

COMPARE the tarantula's two nerve ganglia—the supraesophageal ganglion and the subesophageal ganglion—with the multiple ganglia of a *HONEYBEE* or a *HOUSEFLY*.

CONNECTIONS

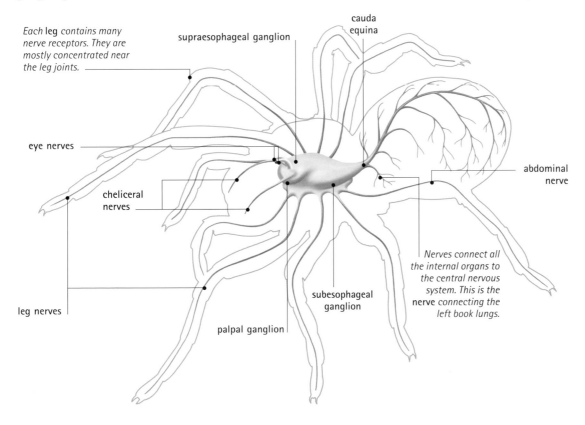

Each leg contains many nerve receptors. They are mostly concentrated near the leg joints.

cauda equina

supraesophageal ganglion

eye nerves

cheliceral nerves

leg nerves

palpal ganglion

subesophageal ganglion

Nerves connect all the internal organs to the central nervous system. This is the nerve connecting the left book lungs.

abdominal nerve

◀ *The tarantula's "brain" is made up of two ganglia: the supraesophageal ganglion and the subesophageal ganglion. Nerve fibers branch from these ganglia to all parts of the spider's body.*

COMPARATIVE ANATOMY

Eyes

Jumping spiders of the family Salticidae use their excellent eyesight to actively hunt prey. For these spiders, vision is crucial. They have eight eyes, with two pairs facing forward and one pair on either side of the prosoma. In contrast, tarantulas have three or four pairs of eyes situated on the front of their prosoma.

▶ *The small eyes of this Mexican red-kneed tarantula are clearly visible on the top of the prosoma. However, this spider relies on detecting vibrations and sounds of prey when hunting.*

▼ TRICHOBOTHRIUM

When the shaft of the trichobothrium is moved—even very slightly—the helmet structure pushes against the dendritic terminals. The movement produces nerve impulses that pass along the dendrites to the brain.

aware if a predator or prey is moving on the ground or even flying close by. Tarantulas are famously hairy, and this appearance is not just for show. The hairs are innervated (they have nerves running along their length), so they provide tactile information. A smaller number of finer hairs called trichobothria are present especially on the lower limbs; trichobothria are extremely sensitive. Research has shown that at least some spiders are able to detect the wing movements of a fly 1 foot (30 cm) away.

A tarantula has slit sense organs all over the outside of its body, but they are most common on the lower segments of the limbs. The slit organs sense sound, vibration, and gravity. Some occur singly, and others are arranged in parallel and are called lyriform organs. Each slit (typically only 1-2 micrometers wide) is covered by a thin membrane, through which a nerve runs. When a spider moves, stresses act on its cuticle, and these narrow or widen the sensory slits. In turn, the change in width of the slit registers with the nerve, giving the spider an appreciation of how its body is moving.

A tarantula also has sensors inside its body. Internal joint receptors in the limbs are clusters of sensory nerve cells, or ganglia, that convey information about the position and movement of the legs. They provide the spider with a picture of the direction of change of a joint and the speed at which it is happening.

There are several groups of other receptors: tarsal organs are small pits on the dorsal surface of each tarsus that probably provide information about humidity; chemoreceptive hairs on the tarsi of the first pair of legs are important for food and reproduction; and thermoreceptors are probably located on the tips of the legs and on spinnerets.

There are up to four pairs of simple eyes, each with a single lens. Spiders have primary and secondary eyes. In primary eyes, the rhabdomeres, or light-sensitive parts of the visual cells in the retina, face toward the light. However, in secondary eyes the rhabdomeres face away, as photoreceptors do in humans.

A **trichobothrium** *is a very fine, innervated (nerve-containing) hair. Trichobothria are much less common than the ordinary hairs on a tarantula's body and are extremely sensitive to air movements.*

shaft

articulating membrane

socket

exoskeleton

dendritic terminal

helmet structure

receptor's lymph cavity

dendrites

sheath cells

Circulatory and respiratory systems

A tarantula's muscles and organs require oxygen in order to function. Oxygen is carried around the spider's body in large protein molecules called hemocyanin. Hemocyanin is the equivalent of human hemoglobin, although it is not as efficient. Hemoglobin can carry 12.8 cubic inches (210 cm³) of oxygen per liter of blood, but hemocyanin can carry less than $\frac{1}{10}$ of this amount.

As in other spiders, tarantulas have a partly open circulation; their circulatory fluid, or hemolymph, is pumped through arteries and veins by a heart, but it is not always contained within vessels. Instead, hemolymph can fill the spider's body cavities, bathing the internal organs. The heart is a tube in the dorsal part of the opisthosoma with several ostia, or tiny slits. The hemolymph enters the heart through the ostia from a membranous bag called a pericardium. When the heart muscles contract, the ostia close and hemolymph is forced from the heart through two arteries. The anterior aorta carries hemolymph into the prosoma, and the posterior aorta takes hemolymph around the opisthosoma. Valves at either end of the heart ensure that blood flows out and not back in. The arteries branch into smaller vessels, and they are open-ended, allowing hemolymph to reach the spider's tissues. After

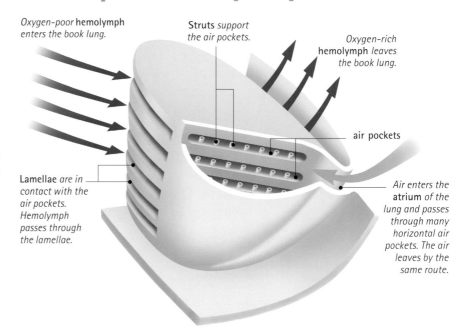

Oxygen-poor hemolymph enters the book lung.

Struts support the air pockets.

Oxygen-rich hemolymph *leaves the book lung.*

air pockets

Lamellae *are in contact with the air pockets. Hemolymph passes through the lamellae.*

Air enters the atrium *of the lung and passes through many horizontal air pockets. The air leaves by the same route.*

▲ BOOK LUNG

Gas exchange takes place in the book lungs. Tarantulas in the genus Eurypelma *have four book lungs, which have a total area of 10.85 square inches (70 cm²).*

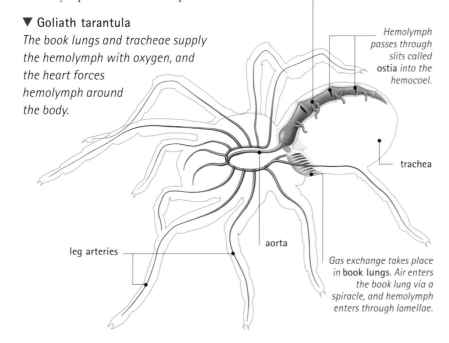

▼ Goliath tarantula

The book lungs and tracheae supply the hemolymph with oxygen, and the heart forces hemolymph around the body.

heart

Hemolymph passes through slits called ostia *into the hemocoel.*

trachea

leg arteries

aorta

Gas exchange takes place in book lungs. *Air enters the book lung via a spiracle, and hemolymph enters through lamellae.*

bathing the inside of the tarantula's body, the hemolymph drains to lacunae, or spaces, in the ventral opisthosoma.

Air is drawn into a tarantula's abdomen through four slits, which lead to breathing organs called book lungs. Hemolymph then passes through the tarantula's book lungs, where gas exchange occurs. Most spiders have one pair of book lungs and one pair of tracheae that branch out throughout the body. However, tarantulas have two pairs of book lungs. Biologists consider this arrangement to be a feature shared with ancient spiders. A book lung is a chamber across which there are numerous sheets of membrane-covered tissue—the pages of the "book"—separated by air spaces. The arrangement provides a large surface area for gas exchange to take place: up to 4.3 cubic inches (70 cm²) in large tarantulas. The two pairs of book lungs in tarantulas serve two pathways of hemolymph circulation: hemolymph from the prosoma passes through the front pair of book lungs, and hemolymph from the opisthosoma flows through the rear pair. The circulatory fluid flows through the sheets, and gas exchange occurs across the membranes. The oxygen-rich hemolymph then flows through veins to the pericardium, from which the hemolymph enters the heart via the ostia. The whole cycle begins again.

Digestive and excretory systems

COMPARE
digestion in the tarantula with that in another arachnid, the *SCORPION*. Both types of arachnids predigest prey before swallowing it as a pulp, and absorption occurs in the midgut.

Like all spiders, tarantulas are carnivores. Some of the larger species prey on frogs, lizards, mice, or small birds. Tarantulas paralyze their prey with a venomous bite delivered by their cheliceral fangs. Spiders have poor internal digestion, so they can take in food only in a liquid form. Thus a tarantula begins the process of digestion before its meal is swallowed. The process is called predigestion. The spider vomits digestive fluids onto the food from its intestinal tract. A combination of manipulation of the prey with the tarantula's pedipalps and with the tiny teeth on its chelicerae, and the operation of the digestive juices, converts even relatively large items into mush. Within a few hours, a captured frog may be transformed into an unrecognizable mess with just a few small bones visible. The venom used to paralyze the prey plays an insignificant role, if any, in predigestion.

Experiments have shown that tarantulas and other spiders can "taste" their food before it is ingested, so they will accept some and reject some. Chemoreceptors must guide them,

but zoologists do not fully understand the processes involved.

From mouth to sucking stomach

The next stage of the digestive process is for the tarantula to get the food into its digestive tract. Food moves from the mouth into the pharynx, esophagus, and sucking stomach, which are all in the prosoma. Both the pharynx and the sucking stomach are lined with muscles. When the muscles contract, the walls of the stomach and pharynx rapidly expand and create a vacuum. The stomach becomes a suction pump. Surprisingly quickly, mashed prey can be sucked through the mouth and pharynx into the sucking stomach. Items of food too large to be digested are filtered out by bristles around the spider's mouth or by the cuticular lining of the rostrum, or labrum (the tarantula's "upper lip"). Regulatory valves ensure that food does not move back out through the pharynx and mouth. Absorption of digested food cannot take place through the walls of the pharynx

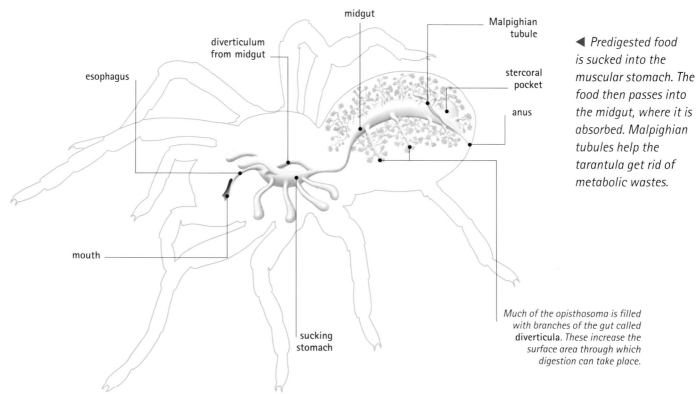

esophagus

diverticulum from midgut

midgut

Malpighian tubule

stercoral pocket

anus

mouth

sucking stomach

◀ *Predigested food is sucked into the muscular stomach. The food then passes into the midgut, where it is absorbed. Malpighian tubules help the tarantula get rid of metabolic wastes.*

Much of the opisthosoma is filled with branches of the gut called **diverticula.** *These increase the surface area through which digestion can take place.*

Paralyzing prey

Most spiders, including tarantulas, use venom to paralyze or kill their prey. Tarantulas have a pair of poison glands—long, cylindrical chambers with a duct that runs to the end of each cheliceral fang. The chelicerae of tarantulas strike downward, but those of most other spiders operate with a sideways pincer action.

A spiral of muscle runs around each poison gland, and when the muscle contracts venom is expelled very rapidly along the ducts. In large tarantulas, the poison glands are small and are contained within the chelicerae; in some other spiders, they are much larger and extend into the prosoma. The chemical composition of spider venom is very variable; it is only rarely dangerous to humans and other large vertebrates but can immobilize smaller prey. For example, robustotoxin in the venom of the funnel-web spider affects the transmission of nerve impulses.

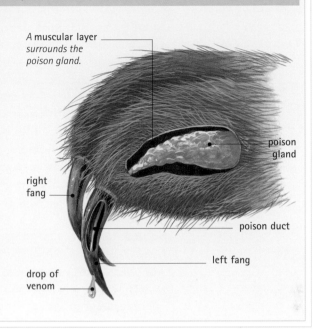

A muscular layer surrounds the poison gland.

poison gland

right fang

poison duct

left fang

drop of venom

▶ POISON GLAND AND FANG

There is a poison gland in each chelicera. Muscles around the poison gland contract rapidly to expel venom along a narrow duct running through the fang. The venom leaves the fang near the tip and is delivered to prey.

and stomach, because they are lined with cuticle. However, beyond the stomach is the midgut, and that is where absorption occurs.

The front part of the midgut is in the prosoma, and from this run two extensions, or diverticula. The diverticula split and branch into many areas of the prosoma, even into the coxae of the walking legs. Other branches of the gut fill much of the interior of the opisthosoma. The extensive nature of these diverticula explains why the spider can digest a large volume of food quickly and why a well-fed spider may not need to feed again for weeks or even months. Waste products in the hemolymph are absorbed by the largest cells in the tarantula's body—nephrocytes—and are then passed into Malpighian tubules, which in turn empty into the stercoral pocket and out through the anus. In addition, mygalomorph tarantulas are able to excrete through paired coxal glands on the coxae, the section of the leg closest to the body.

◀ *The king baboon tarantula is a burrowing species that lives in eastern Africa. It is believed to have some of the most toxic venom of any tarantula and preys on invertebrates and small reptiles and mammals.*

Reproductive system

COMPARE the
tarantula's method
of sperm transfer
with that of a
SCORPION. A male
tarantula passes
sperm directly into
the female's body,
whereas a male
scorpion deposits
sperm on the
ground for the
female to pick up.

Male spiders are almost always smaller than females. Because of this size difference, called sexual dimorphism, males mature more quickly than females. Once sexually mature, males wander in search of females, and—if they are successful—perform some kind of courtship ritual before mating. The ritual is important, because females respond only to the mating gestures of their own species. This behavior minimizes the number of wasted matings and ensures that males are not mistaken for prey.

Sexual organs

Male tarantulas produce sperm in paired testes. The sperm pass along ducts that meet before leaving the body through a single genital opening. In some species the sperm cells are packaged together. Spiders do not transfer sperm directly from genitals to genitals when

CLOSE-UP

Silk production

All spiders possess spinning glands, from which silk is excreted. Male tarantulas produce silk to make a web on which to deposit their sperm. Female tarantulas excrete silk to make a protective cocoon for themselves and their eggs. Other spiders snare prey with silken webs. The silk glands are in the abdomen and lead to three pairs of spinnerets. Silk is secreted in the form of tiny protein droplets. The transition from liquid silk to a solid thread occurs when tension orients the molecules so they are parallel to one another. Spider silk is both very strong and very elastic.

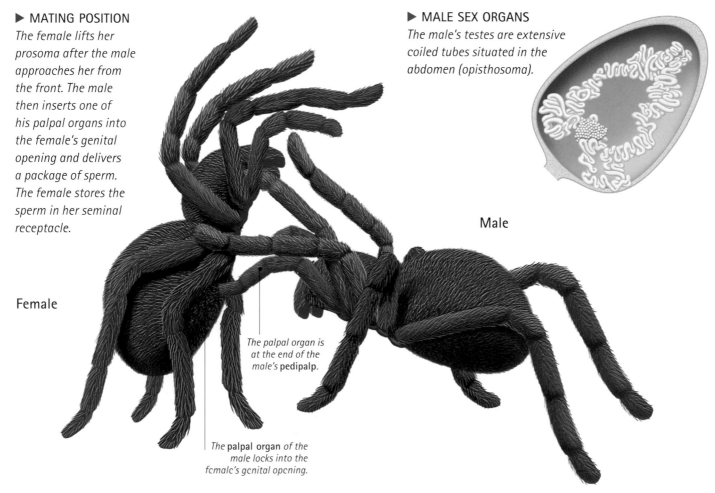

▶ MATING POSITION
The female lifts her prosoma after the male approaches her from the front. The male then inserts one of his palpal organs into the female's genital opening and delivers a package of sperm. The female stores the sperm in her seminal receptacle.

▶ MALE SEX ORGANS
The male's testes are extensive coiled tubes situated in the abdomen (opisthosoma).

Male

Female

The palpal organ is at the end of the male's **pedipalp.**

The **palpal organ** *of the male locks into the female's genital opening.*

they mate. Instead, the male spins a web beneath its genital opening to catch the sperm. The male then dips its pedipalps into the web to pick up the sperm, a process called palpal lubrication. The sperm is stored in a bulb, or palpal organ, at the tip of the pedipalps, ready for mating.

When mating, a male tarantula locks its palpal organ with the female's genital opening. The fit is perfect in males and females of the same species, but males and females of different

▼ *A female goliath tarantula with some of her spiderlings. The female lays 100 to 400 eggs, from which the spiderlings hatch after six to eight weeks.*

IN FOCUS

Approaching a female

The females of many tarantula species dig deep burrows and line them with silk. Mature male tarantulas react if they encounter the pheromone-laden silk surrounding a female's burrow. Males in the genus *Aphonopelma*, for example, tap out a greeting on the threshhold of the burrow. The female emerges to view the visitor, and if she is responsive she will tilt back onto her hind legs, hold her front legs up, and bare her fangs. The male then advances and grasps the female's fangs with his pedipalps. The male then lets go of the female's fangs and transfers sperm from his palpal organ to her genital opening. Many male tarantulas also perform a quivering display when approaching the female, probably to indicate that they are ready to mate.

species do not match. When the male's palpal organ is locked to the female's genital opening, sperm passes into the female's seminal receptacle. Female tarantulas, like other spiders, produce very large numbers of eggs in paired ovaries. The eggs pass along oviducts and are fertilized by sperm in the uterus externus.

The female lays up to 1,000 fertilized eggs, one by one. Spiders in the genus *Cupiennius* (not tarantulas) lay up to 2,000 eggs in eight minutes. Tarantula eggs are laid in a silk cocoon in the female's burrow. The cocoon provides a barrier against egg parasites and insulation against changes in temperature. The female guards the eggs until tiny spiderlings hatch between six and nine weeks later. After another two or three weeks, the spiderlings leave the safety of the burrow.

TIM HARRIS

FURTHER READING AND RESEARCH

Jackson, T,. and J. Martin (eds). 2003. *Insects and Spiders of the World*. Marshall Cavendish: Tarrytown, NY.

Ruppert, Edward E., and Robert D. Barnes. 1994. *Invertebrate Zoology*. Saunders College: Fort Worth, TX.

Tortoise

CLASS: Reptilia ORDER: Chelonia FAMILY: Testudinidae

Tortoises occur mainly in tropical and subtropical regions and on all major landmasses except Antarctica and Australia. They also live on some groups of small islands, notably the Galápagos and the Mascarene Islands (Mauritius, Aldabra, and neighboring islands). Members of the genus *Geochelone* have a widespread distribution, from the Galápagos and mainland South America through Africa and into southern Asia. Eight species are endangered, one critically so, and another 15 species are thought to be vulnerable. There are almost no species of tortoises whose numbers have not been reduced significantly in recent years, by some combination of habitat destruction, human predation on adults or eggs, predation and competition by introduced animals, and collection for the pet trade.

▼ *There are around 12 genera and 40 species of living tortoises in the family Testudinidae. However, scientists do not agree on how to classify all the tortoise genera.*

Anatomy and taxonomy

Scientists group all organisms into taxonomic groups based largely on anatomical characteristics. Tortoises are reptiles. They belong to the order Chelonia (sometimes called Testudinata), which includes the freshwater and marine turtles and contains about 400 species all together, divided into 13 families. The living tortoises are placed in the family Testudinidae, with 12 genera and 47 species.

- **Animals** Animals are complex, multicellular organisms that rely on other organisms for food. Most are capable of movement (by using muscles), and they have sense organs that enable them to detect external stimuli and respond if necessary.

- **Chordates** At some stage during their development these animals have a stiff supporting rod called a notochord along their back.

- **Vertebrates** Vertebrates are animals with a backbone, or spine, consisting of connected bones called vertebrae. The head, trunk, and limbs are attached to the backbone by a series of muscles that control movement. Vertebrate muscles are bilaterally symmetrical about the skeletal axis: the muscles on one side of the backbone are the mirror image of those on the other side.

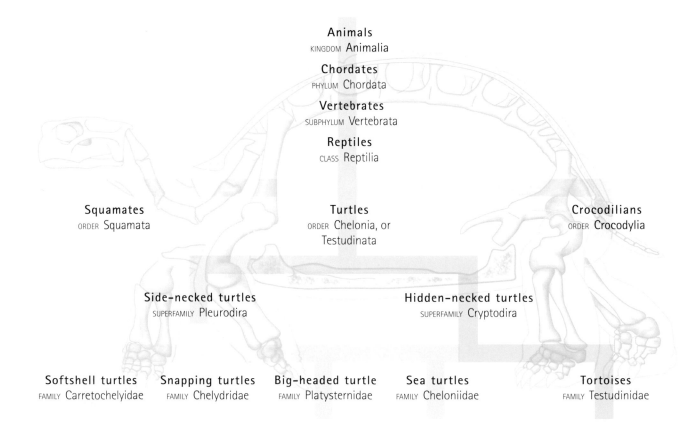

Animals
KINGDOM Animalia

Chordates
PHYLUM Chordata

Vertebrates
SUBPHYLUM Vertebrata

Reptiles
CLASS Reptilia

Squamates
ORDER Squamata

Turtles
ORDER Chelonia, or Testudinata

Crocodilians
ORDER Crocodylia

Side-necked turtles
SUPERFAMILY Pleurodira

Hidden-necked turtles
SUPERFAMILY Cryptodira

Softshell turtles
FAMILY Carretochelyidae

Snapping turtles
FAMILY Chelydridae

Big-headed turtle
FAMILY Platysternidae

Sea turtles
FAMILY Cheloniidae

Tortoises
FAMILY Testudinidae

● **Reptiles** These are a group of four-legged vertebrates, although some have lost their legs during the course of evolution (and a few have lost two and kept two). They have thick, horny skin that is usually divided into plates called scales. Most reptiles lay eggs with a waterproof shell, but females of some species retain their eggs inside their bodies until the eggs are ready to hatch. Many snakes and lizards are viviparous: they give birth to living young. A few snakes and lizards have a simple kind of placenta through which the mother and embryo exchange nutrients and expel waste products. Reptiles are not endothermic (warm-blooded) animals like birds and mammals, but most can control their body temperature by moving into hot places when they need to warm up and into cooler places when they need to lose heat. A reptile that has been basking for some time may be warmer than a bird or mammal.

● **Turtles and tortoises** Turtles may live in freshwater or salt water, or they may live on land, in which case they are usually called tortoises. Together, tortoises and turtles are sometimes called chelonians. All have a bony shell covering their body: the domed upper shell is called the carapace, and the shell beneath the tortoise is called the plastron. In some highly aquatic species such as the marine turtles, snapping turtles, and softshell turtles, the shell is reduced so the turtle cannot withdraw its head and limbs completely. In most species, however, the shell is the first line of defense.

● **Land tortoises** Tortoises are characterized by a heavily built, domed shell. Their front legs are slightly flattened and their back legs are cylindrical and short, like an elephant's legs. They do not have separate toes, but each leg ends in a

▲ *The Aldabra giant tortoise originates from the Aldabra Atoll in the Seychelles in the Indian Ocean and rivals the Galápagos giant tortoise in weight and size.*

number of claws. Their shell ranges in length from 3.5 to 55 inches (9–140 cm). Tortoises are not brightly colored, but some, such as the South African tent tortoise, the Madagascar radiated tortoise, and the Indian starred tortoise, have patterned shells. All tortoises eat plant material, and many eat a small amount of animal material as well, including carrion.

FEATURED SYSTEMS

EXTERNAL ANATOMY Tortoises are distinctive animals with bony shells covering their upper and lower surfaces. The head and front limbs can be withdrawn into an opening of the shell at the front, and the hind limbs and tail can be withdrawn into an opening at the rear. This is an excellent means of defense. *See pages 1268–1271.*

SKELETAL SYSTEM Tortoises' skeletons are highly modified to fit into their shell, with some bones fused to the inside of the shell. The pectoral girdle is positioned inside the rib cage, a feature that is unique among vertebrates. *See page 1272.*

MUSCULAR SYSTEM Tortoises' movements are limited, owing to the constraints placed on them by a bony shell, and their muscular system is therefore in some respects simplified. *See page 1273.*

NERVOUS SYSTEM Tortoises appear to have a highly developed sense of vision and can distinguish certain

colors. They use sight for navigation and for searching for food. Their other senses include smell and taste, used mainly for finding food and also for communication among individuals. *See page 1274.*

CIRCULATORY AND RESPIRATORY SYSTEMS Tortoises have a three-chamber heart, although there is some separation of oxygenated and deoxygenated blood in the single ventricle. *See pages 1275–1276.*

DIGESTIVE AND EXCRETORY SYSTEMS Most tortoises are almost exclusively herbivorous and have a long gut that allows plenty of time for food to be digested as it passes through. They have no teeth but use their sharp beaklike jaws to tear off pieces of plant material. *See page 1277.*

REPRODUCTIVE SYSTEM All tortoises lay eggs. Mating can be a troublesome affair, with the male attempting to balance himself on the domed carapace of the female before he can attempt copulation. *See pages 1278–1279.*

External anatomy

CONNECTIONS

COMPARE the shell shape of a tortoise with that of an aquatic **SNAPPING TURTLE**. The environment in which turtles live is often obvious from the shape of their shell. Land-living tortoises have a high, domed shell, whereas aquatic turtles have a low, streamlined shell.

The general body shape of tortoises and turtles has not changed significantly from that of the earliest turtles, which lived more than 200 million years ago. The land tortoises, in particular, are easily recognizable, and it is difficult to confuse then with any other group of animals, apart from a few of the more terrestrial turtles, such as box turtles in the genus *Terrapene*. Land tortoises have a bony shell consisting of a domed top shell, or carapace; and a flattish shell, or plastron, covering the underside. The carapace and plastron are joined at each side by a section of shell called the bridge.

The carapace is made up of about 50 individual plates that have their origins in the dermal (inner) layer of the skin. The plates are joined together with sutures; these are areas where the edges of adjacent plates interlock with each other. Each plate is named according to its shape and position, and these are constant in any given species although they may change shape slightly throughout an individual's life. The plastron is made of 11 such bony plates.

CLOSE-UP

Shells with hinges

Members of several families of turtles have a hinged plastron, which allows them to close up tightly when they have withdrawn into their shell. The most familiar of these are the box turtles, of which there are four species living in North America and Central America. Some of the land tortoises, including the Mediterranean tortoises and the two species of *Pyxis* tortoises from Madagascar, have similar arrangements. The five species of land tortoises of the genus *Kinixys*, however, are unique because they have a hinged carapace. Called hinge-back tortoises, they occur in southern and eastern Africa.

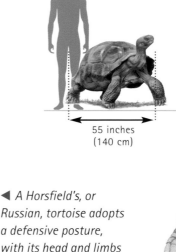

55 inches
(140 cm)

◀ A Horsfield's, or Russian, tortoise adopts a defensive posture, with its head and limbs tucked between the carapace and the plastron. The carapace of this species is almost as broad as it is long. Females grow larger than males, up to 9 inches (22 cm) long.

The **hind legs** *are very robust.*

The bony plates are overlaid by thinner horny plates, originating in the epidermal (outer) layer of the skin. There are 38 or more of these on the carapace and 12 on the plastron. The outer plates are usually a shade of brown or buff and may be plain, spotted, or marked with radiating lines; the plates give each species its characteristic markings. Young tortoises are often more brightly marked than adults.

▼ Galápagos giant tortoise
This huge species grows up to 4 feet (1.2 m) long and may weigh 500 pounds (215 kg). It has a domed carapace, pillarlike legs, and a relatively small head.

The carapace, or upper shell, is domed and very large. Some Galápagos giant tortoises have a less domed carapace.

nostrils

Tortoises have a relatively small head. They do not have teeth. The animals tear off and grind up food with their horny beak.

The long neck allows the tortoise to reach up or down for plant material.

A tough plastron, or lower shell, protects the underside of the body.

The forelegs are protected by bony scales.

Tortoises do not have webbed feet. Instead, they are armed with strong claws.

1269

Shell shapes in the Galápagos tortoises

Fourteen subspecies of giant tortoises evolved and lived on the Galápagos archipelago, although three of these became extinct in recent times. All are descended from the same ancestor, but the shells of populations living on separate islands are shaped differently. As long ago as the early 19th century, naturalists noticed that tortoises living on the drier islands had shells that are shaped like saddles, with a flared opening at the front. Careful observation showed that the saddle-backed subspecies browsed on low-growing plants, especially prickly pear cacti. Over the millennia, natural selection had favored those tortoises that could reach up highest, and so the shells of these populations had gradually changed shape over many generations. Tortoises from wetter islands had typical dome-shape shells because they grazed on low vegetation and there was no selective pressure on them to evolve the flared shells.

▼ **Dome-backed Galápagos giant tortoise**
This is the shape of carapace typical of tortoises on the larger Galápagos islands, where there is more low-growing vegetation.

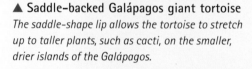

▲ **Saddle-backed Galápagos giant tortoise**
The saddle-shape lip allows the tortoise to stretch up to taller plants, such as cacti, on the smaller, drier islands of the Galápagos.

◄ **Seychelles giant tortoise**
This form is intermediate between the two extreme shapes of the Galápagos giant tortoises.

Although the shell protects the tortoise, it also limits movement, and land tortoises are slow-moving, rather ponderous animals. In large species, the shell is reduced in thickness; otherwise, it would be too heavy for the animal to lift. The rounded shape of the shell makes it difficult for predators to grip in their jaws. In contrast, aquatic turtles have a much lower carapace and are relatively streamlined. They can move surprisingly quickly through the water, and some species of aquatic turtles have greatly reduced shells.

Tortoises' limbs, tail, and neck are covered with scales of different sizes. Their front feet are flattened, whereas their hind feet are club-shaped. Each foot ends in three or five claws, but there are no separate toes. The feet of highly aquatic species, such as the marine turtles, are modified into flippers, and those of some freshwater turtles are webbed. All species

▲ *These two male Galápagos giant tortoises are competing for the right to mate with a female. They stretch their necks, ram each other, and produce a loud roar to establish which animal is dominant.*

have a beaklike mouth with a sharp cutting edge for biting off pieces of food. In the case of land tortoises, this food is usually some type of vegetation. Tortoises have a short, stubby tail, which serves little or no purpose, although a male's tail sometimes ends in a clawlike scale that helps it grip the carapace of a female during copulation.

▶ *The desert tortoise lives in the deserts of northern Mexico and the southwestern United States. All individuals have a domed carapace, but the plastron varies between the sexes: females have a flat plastron, whereas males have a concave area toward the rear. This feature helps a male balance on a female during mating.*

Skeletal system

CONNECTIONS

COMPARE the backbone of a tortoise with that of a snake such as a *GREEN ANACONDA*. In a tortoise, the backbone lacks flexibility, since it is mostly fused to the inside of the shell, whereas snakes have many highly bendable vertebrae.

COMPARE the two limb girdles of a tortoise (the pelvic and pectoral girdles) with the single pelvic girdle of a *GREEN ANACONDA*.

A tortoise's internal skeleton consists of a skull and a series of vertebrae and ribs, onto which are arranged the pelvic and pectoral girdles and the bones of the limbs. The backbone consists of 40 to 50 vertebrae, but all of these apart from the ones in the neck and tail are fused to the neural plates on the inside of the carapace. The neural plates are those that follow the midline of the carapace from front to back. Similarly, the ribs have become fused to the bones of the shell, and the ribs are associated with the costal plates, which are arranged on both sides of the midline. This arrangement limits the number of movable bones to those forming the neck, tail, limbs, and limb girdles.

The rib cage encloses the shoulder girdle in turtles and tortoises, an arrangement that is unique among vertebrates. The neck bones articulate to allow the tortoise to look around, graze, and—in land tortoises and freshwater turtles with complete shells—to withdraw the head into the safety of the shell. Turtles with a reduced shell, such as snapping turtles and softshells, cannot do this.

IN FOCUS

Pancake tortoises

When the embryo of a tortoise begins to develop a shell, there are spaces between the individual bones of the carapace and usually also between the bones of the plastron. By the time the embryo hatches from the egg, these spaces have closed—except in the East African pancake tortoises. This species lives among rock outcrops, and its shell is flat and flexible, allowing it to wedge itself into crevices. The spaces between the bones of the carapace give the shell flexibility, so when danger threatens, a pancake tortoise can wedge itself into a crevice. Afterward it is almost impossible to dislodge.

Marine turtles also have spaces between some of the bones of the carapace because the costal bones do not reach the peripherals. This arrangement reduces weight and so increases the buoyancy of the turtle.

*A tortoise has no teeth, but its jaws are covered with a horny layer, forming a **beak** with sharp cutting edges. In land tortoises, the upper jaw ends in a hook, helping the tortoise to slice through tough vegetation.*

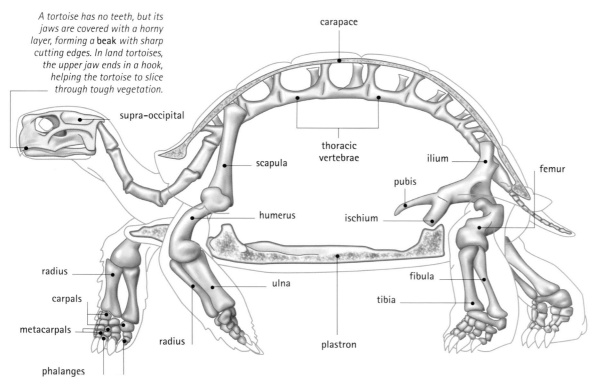

◄ Galápagos giant tortoise
The skull is heavily built and rigid, unlike that of a snake or lizard. Apart from the eye sockets, the skull has no other openings, a condition called anapsid. When threatened, the tortoise can withdraw its head and legs under the enormous carapace.

carapace

supra-occipital

thoracic vertebrae

scapula

ilium

femur

pubis

humerus

ischium

radius

carpals

metacarpals

radius

ulna

fibula

tibia

phalanges

plastron

Muscular system

A tortoise's muscular system is very different from that of other animals because its shell forms a rigid box around most of the body and limits movement to the neck, head, limbs, and tail. The limb girdles are suspended below the fused vertebrae by a series of muscles that act as a sling. Bundles of muscles attached to the limb girdles and the limbs act as extensors and flexors: they allow the tortoise to bend and straighten its limbs. Other muscles act as rotators, allowing the tortoise to turn the limbs.

Leg and neck muscles

Land tortoises are not known for their speedy movement. Smaller species of tortoises often raise themselves well clear of the ground when moving and make progress in a series of jerks or lurches. Larger tortoises may lower their shell to the ground occasionally, presumably to rest their leg muscles. Despite their slow movement, tortoises large and small have plenty of stamina and often cover surprisingly large distances in a short space of time.

As well as functioning in locomotion, the limb flexor muscles enable a tortoise to draw its limbs up into its body when it is protecting itself. Muscles in the neck enable it to withdraw its head. In land tortoises and other members of the suborder Cryptodira (hidden-necked turtles), the head is withdrawn by bending the neck into a vertical S-shaped kink. This action allows the head to move straight back into the safety of the shell. Land tortoises usually use their front feet to cover what little is showing of the head. In contrast, side-necked turtles swing their head to one side by bending their neck in the horizontal plane. Thus their neck is not as well protected as that of hidden-necked turtles. Land tortoises cannot withdraw their tail in the way they withdraw their limbs but simply swing it to one side. In the Mediterranean tortoises (in the genus *Testudo*) and the hinge-backed tortoises (genus *Kinixys*), there are other groups of muscles that open and close the hinged plastron or carapace.

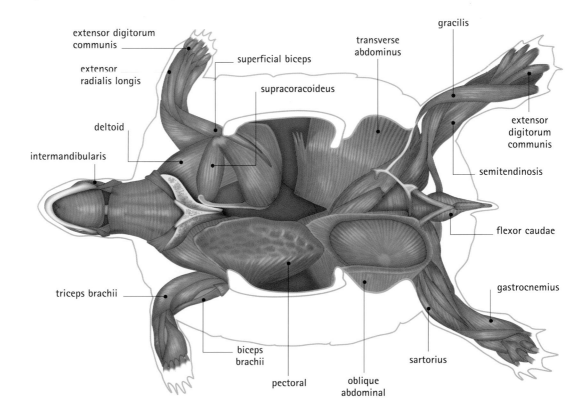

▶ Galápagos giant tortoise

The tortoise's sturdy neck and limbs have well-developed muscles. Within the protective shell, the major muscles are thin sheets of tissue.

extensor digitorum communis

extensor radialis longis

deltoid

intermandibularis

superficial biceps

supracoracoideus

transverse abdominus

gracilis

extensor digitorum communis

semitendinosis

flexor caudae

gastrocnemius

triceps brachii

biceps brachii

pectoral

oblique abdominal

sartorius

Nervous system

CONNECTIONS

COMPARE the sense of taste in a tortoise with that of a *GREEN ANACONDA*. The green anaconda has a vomeronasal organ, also called Jacobson's organ, which allows the snake to "taste" the air.

A tortoise's nervous system is little different from that of other vertebrates. As in other vertebrates, the central nervous system (CNS) consists of the brain and spinal cord. The CNS is connected via the nerves of the peripheral nervous system (PNS) to sensory organs and to responsive structures such as muscles. A tortoise's CNS is more developed than that of an amphibian, and the brain has larger cerebral hemispheres and a differentiated medulla and cortex. The cerebellum is also larger, resulting in a greater ability to coordinate movements.

A tortoise's senses of sight, taste, and smell are probably very good. Laboratory experiments have shown that a tortoise can distinguish certain colors and also distinguish between vertical and horizontal stripes. Galápagos tortoises are attracted by the color yellow, perhaps associating it with the flowers they eat. Members of this species and others often pause to lift their head and scan their surroundings before resuming a journey, indicating good long-distance vision. Marine turtles probably find their way to breeding beaches by the faint outline of coastlines.

CLOSE-UP

Rods and cones

Tortoises, turtles, and crocodilians have rod and cone cells in the retina of the eyes. Rod cells provide monochromatic (black-and-white) vision even in very dim light, whereas cones are responsible for color vision. By contrast, other reptiles have more limited retinas: a "primitive" snake's retina contains only rods; advanced snakes have rods and cones, but the cones are thought to be transformed rods; and lizards have only cones.

Taste and smell are used to find and identify food and to recognize other animals that may be potential mates or enemies. The rhythmic pumping action of a tortoise's throat brings particles through its nostrils into the pharynx, and from there to its internal nostrils. Odors are analyzed in the nostrils, and the animal bases its decisions on the results. A tortoise that recognizes the smell of potential food, for example, will home in on it and, having found it, use the taste cells on the tongue to check the food's edibility.

Aquatic turtles rely more heavily on their senses of smell and taste than do land tortoises because the aquatic reptiles often feed in murky water, where they can see very little. They will gulp water to test for the odor of edible material and sometimes chew on inedible material before spitting it out. Aquatic turtles do use their sense of sight in defense, however, and dive to the bottom at the slightest movement that might indicate a threat.

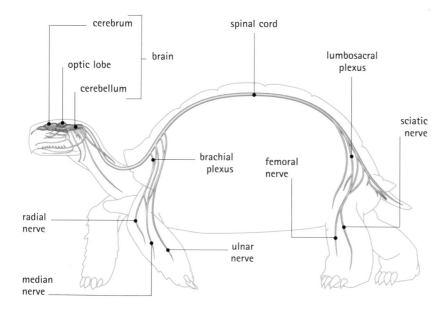

◄ Galápagos giant tortoise
One feature of the nervous system of chelonians is the pineal body, a tiny organ at the top of the cerebrum that detects the presence and intensity of light. The pineal body also secretes the hormone melatonin, but the functions of the organ are not fully understood.

cerebrum
optic lobe
cerebellum
brain
spinal cord
lumbosacral plexus
sciatic nerve
brachial plexus
femoral nerve
radial nerve
ulnar nerve
median nerve

Circulatory and respiratory systems

Tortoises and turtles, unlike most other vertebrates, are unable to move air into and out of their lungs by moving their rib cage, since the ribs are fused to the bones of the shell. Instead, tortoises expand and contract muscles in their legs to pump air into the lungs and then use ventral (underside) muscles to push their intestines up and force the air back out again. Even when a tortoise has retracted its limbs, the slight rhythmic pumping action of its feet is visible.

In addition to their paired lungs, which are similar to those of other vertebrates, tortoises may also have ways of obtaining additional oxygen. Several aquatic species have throat pouches containing many blood capillaries across which gaseous exchange takes place. As the throat is inflated and deflated, water is moved across the surface of the pouch, and oxygen is taken from it in a method reminiscent of gills.

Some aquatic species also have a pair of thin-walled sacs called cloacal bursae (singular, bursa) branching off their cloaca. Again, water is moved in and out of these by rhythmic

IN FOCUS

Heavy breathers

Tortoises and turtles can hold their breath for long periods. There are two reasons for this: they have a low oxygen requirement, and they are very efficient at emptying their lungs and refilling them with oxygen-rich air. In contrast, many mammals are able to exchange only a small proportion of air with each breath. Softshell turtles, which are highly adapted for an aquatic lifestyle, remain on the bottom of ponds, lakes, and rivers and rarely emerge from the water. In shallow water, softshell turtles can continue to breathe by extending their neck and breathing through their nostrils, which are at the tip of their pointed snout, using them like snorkels. In deeper water softshell turtles may rely entirely on underwater respiration, absorbing up to 70 percent of their oxygen through the skin and 30 percent through the throat.

muscular action, and oxygen is extracted via blood vessels in the walls of the sacs. Using these supplementary methods of respiration, both tortoises and turtles can stay underwater for much longer than they could if they had to rely entirely on their lungs.

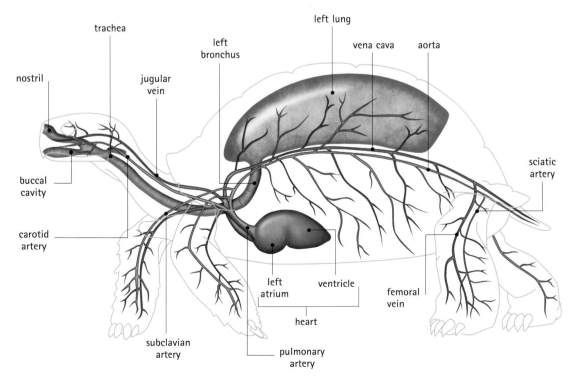

◀ Galápagos giant tortoise
The tortoise has a three-chamber heart that pumps blood to the lungs to pick up oxygen from inhaled air. The blood returns to the heart and is then pumped around the body.

1275

▲ *The green sea turtle is an aquatic species, and it can spend up to six hours underwater. Like other chelonians it is able to breathe in water through the skin of its mouth and throat.*

Blood circulation

A tortoise's heart has only three chambers: two atria and one ventricle. The ventricle is partially divided by an incomplete flap called the septum, which limits the mixing of oxygen-rich arterial and oxygen-poor venous blood. The veins empty venous blood into the right atrium, which then contracts and pumps the blood into the right side of the ventricle. From there, blood enters the pulmonary artery and then goes to the lungs, which remove carbon dioxide and replace it with oxygen. The oxygen-rich blood flows back to the heart, where it enters the left part of the ventricle and is pumped out again, this time through the dorsal aorta and around the rest of the tortoise's body.

Thermoregulation

Like all reptiles, tortoises are cold-blooded, or endothermic: they are unable to produce body heat from metabolism and must rely on outside sources. This does not mean, however, that tortoises are always colder than birds and mammals. In cool weather, tortoises bask in the sun until they have raised their temperature to a preferred level. Then they go off in search of food. If their search for food takes the tortoises to a cooler place, they may return to an open area again to top up their body temperature from time to time. During the winter months, the air may rarely or never reach a suitable temperature, and then they will remain in a sheltered place until the spring. Some species hibernate in the mud and silt at the bottom of a pond. Because their metabolism is reduced almost to a standstill, they can survive for several weeks or months without coming to the surface for air.

At the other extreme, tortoises must avoid becoming too hot, so species that live in tropical climates often shelter in the shade or in burrows during the hottest parts of the day. Some species, such as the Galápagos tortoise, enter shallow lakes and pools during midday to keep cool. Aquatic species may burrow into the mud at the bottom of a pool if prolonged drought causes the water to dry up.

Digestive and excretory systems

Land tortoises eat mostly plant matter. Since they have a heavy shell and can move only ponderously, they cannot prey on any but the slowest-moving animals. Tortoises do not have teeth but use their sharp, beaklike jaws to bite off pieces of leaves, flowers, and grass, sometimes using their front feet to hold the item still. Tortoises have a small, rounded tongue, which is attached to the floor of the mouth and kept moist with saliva. They use the tongue to manipulate pieces of food once the food is in the mouth. Food passes down the short esophagus to the stomach, which has thin walls and many glands that secrete digestive chemicals.

Pyloric and ileocecal valves

The intestine is divided into two parts by a pair of valves: one at the entrance to the small intestine (the pyloric valve); and the other at the junction of the small and large intestines (the ileocecal valve). Digestion is slow and depends largely on the body temperature and diet of the tortoise. Chelonians (mainly freshwater turtles) that eat animal material and that live in warm places digest their food fairly quickly, whereas digestion takes longer in species that eat plant matter and inhabit relatively cool places. The undigested remains of food, along with other waste material, are voided from the body at the cloaca.

Tortoise intestines

Large plant-eating tortoises and turtles have an especially long intestine and digest their food very slowly: green turtles typically take 7 days to process their food (seaweeds), and the Aldabra giant tortoise takes almost 10 days. In contrast, pond turtles, which eat mainly insects and other small animals, hold food in their esophagus for only 3 to 4 hours, after which it spends a further 8 to 10 hours in the stomach and then takes a final 20 hours to move through the intestines. Thus the total time taken for a food item to pass through a pond turtle's body is 30 to 35 hours.

The excretory system of tortoises and turtles consists of a pair of kidneys attached to a stretchable bladder positioned near the cloaca. In most chelonians the waste by-products of metabolism are diluted with water and carried out of the body as urine. Land tortoises have a slightly modified system, however, especially if they live in dry environments. These tortoises convert their waste products into uric acid, a white crystalline substance that requires very little water to carry it out of the body. This arrangement conserves precious water in an environment where water is often in short supply. Snakes and some species of lizards also void uric acid rather than liquid urine, for the same reason.

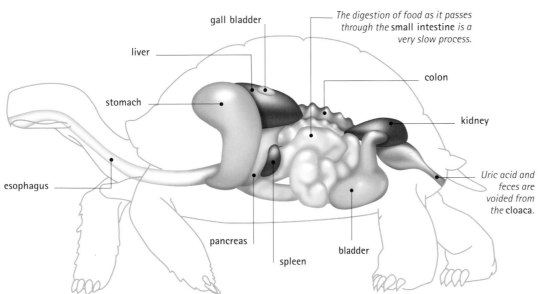

gall bladder

The digestion of food as it passes through the small intestine is a very slow process.

liver

colon

stomach

kidney

esophagus

Uric acid and feces are voided from the cloaca.

pancreas

bladder

spleen

◀ Galápagos giant tortoise
Plant-eating tortoises have a very long digestive tract, and the digestive process takes several days. Metabolic wastes are filtered from the blood by a pair of kidneys and excreted as uric acid.

Reproductive system

COMPARE the internal fertilization of a tortoise with the very different internal fertilization of a *NEWT* and with the external fertilization of a *BULLFROG*.

Tortoises have internal fertilization, unlike most frogs, for example, in which fertilization is external. A male tortoise has a single penis (snakes have a pair of hemipenises) attached to the inside wall of the cloaca. The penis is connected to the testes by a pair of tubes, or vasa deferentia (singular, vas deferens), through which the sperm travel. The female reproductive organs include a pair of ovaries; these are connected by an oviduct to a combined excretory and reproductive opening called the cloaca. Fertilization takes place in the upper part of the oviduct. A female tortoise can produce fertile eggs for several years after a single mating by using sperm stored in the oviduct, although the fertility of the sperm gradually diminishes over time.

All turtles lay eggs. They may be spherical (usually in the large species) or oval. The egg-shells may be soft and leathery or hard and brittle, depending on the species. Some hard-shelled eggs are soft when they are first laid, enabling them to pass through the female's cloaca comfortably. Soft-shelled eggs swell slightly during incubation. The shells are formed in the oviduct, after the egg is fertilized. First, the egg is coated with a layer of the protein albumin; this is followed by several layers of protein fibers impregnated with crystals of argonite, which are produced in endometrial glands lining the oviduct.

Land tortoises lay hard-shelled eggs. These are placed in a small flask-shape pit that that the tortoises dig in soil and then cover over. The eggs are incubated by the warmth of the sun and typically take 100 to 150 days to hatch, although some take longer, especially in cooler climates. Leopard tortoise eggs, for example, can take up to 18 months to hatch under some circumstances. The number of eggs laid depends on the species and the size of the individual female. The pancake tortoise always

▶ Galápagos giant tortoise
The male tortoise has two testes, which produce sperm, and a penis inside the cloaca. The female has two ovaries, which produce eggs, and a cloaca.

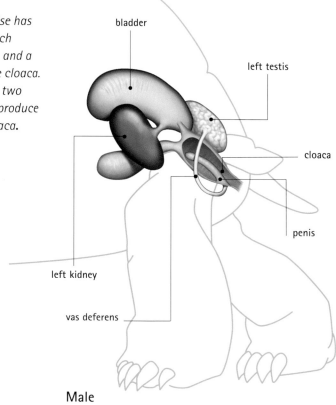

bladder

left testis

cloaca

penis

left kidney

vas deferens

Male

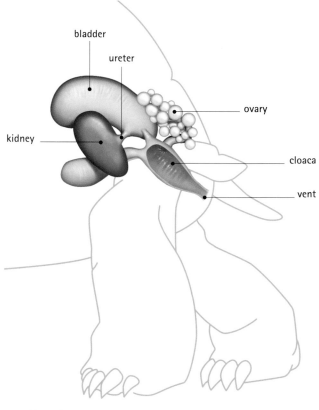

bladder

ureter

ovary

kidney

cloaca

vent

Female

◄ *These two spurred tortoises are mating. The male circles the female several times before climbing onto her carapace and inserting his penis into her cloaca. After her eggs are fertilized, the female digs a nest hole and lays 15 to 30 eggs. The eggs are covered over and hatch about eight months later.*

lays a single egg, probably because its flat body allows only a limited amount of space for the development of the egg. The smallest tortoise, the speckled padloper, lays one to three eggs. In contrast, marine turtles frequently lay more than 100 and sometimes more than 200 eggs. In such cases, several smaller clutches are laid a few days apart over a week or more. Even very large land tortoises cannot come close to matching the quantity of eggs laid by turtles. The Galápagos giant tortoise, for example, lays only 4 to 10 eggs; the giant Aldabra tortoise lays 4 to 14 eggs; and the very rare Madagascar plowshare tortoise lays up to 6 clutches totaling no more than 20 eggs.

Because of the domed shape of the shell, a male tortoise may struggle to mount a female during mating. However, the task is made a little easier by the shape of the male's plastron, which is slightly concave. Males of some species are significantly smaller than females, although this difference is more common among freshwater turtles than among land tortoises: male map turtles may be less than 25 percent of the weight of the female of the species.

CHRIS MATTISON

COMPARATIVE ANATOMY

Internal and external fertilization

There is a fundamental difference between fertilization in reptiles and in amphibians. Like all reptile eggs, tortoises' eggs are fertilized while they are in the oviduct, usually shortly after the male has transferred sperm to the female during copulation. Bullfrogs' eggs, like those of most other frogs, are fertilized externally after the female has expelled them. Instead of transferring his sperm directly into the female's oviduct, a male bullfrog releases his sperm into the water. Although there is a mating embrace, called amplexus, there is no direct sperm transfer from the male to the female, and male frogs do not have a penis.

In newts, fertilization is internal, as it is in tortoises, but the male does not introduce the sperm directly into the female's reproductive tract. Instead, the male simply places the sperm on the ground (or the bottom of a stream or pond) and the female picks it up, usually after an elaborate courtship dance. Internal fertilization is a more certain way of ensuring fertilization, although amphibians produce many more eggs and can afford to waste some of them.

FURTHER READING AND RESEARCH

Halliday, T., and K. Adler. 2002. *The New Encyclopedia of Reptiles and Amphibians*. Firefly: Toronto.

Harris, T. (ed.). 2003. *Reptiles and Amphibians*. Marshall Cavendish: Tarrytown, NY.

Orenstein, R., G. Zug, and J. Mortimer. 2001. *Turtles, Tortoises, and Terrapins*. Firefly: Toronto.

Trout

CLASS: Osteichthyes ORDER: Salmoniformes
FAMILY: Salmonidae GENUS: *Oncorhynchus*

The rainbow trout is probably the most widely distributed freshwater fish. Its original range is eastern Russia and western North America, but it has been introduced into many other countries. It is a cool-water fish, favoring well-oxygenated water that does not rise above 59°F (15°C) in summer. The species migrates between freshwater and seawater, and its body readily adapts to these different conditions. Adults feed on a wide range of aquatic animal life, ranging from insects and mollusks to crustaceans and fish.

Anatomy and taxonomy
Scientists categorize all organisms into taxonomic groups based partly on anatomical features. The rainbow trout is one of 10 species of Pacific salmon and trout.

● **Animals** These organisms are multicellular and gain their food by consuming other organisms. Animals differ from other multicellular life-forms in their ability to move from one place to another (in most cases, by using muscles). They generally react rapidly to touch, light, and other stimuli.

● **Chordates** At some time in its life cycle, a chordate has a stiff, dorsal (back) supporting rod called the notochord, which runs all or most of the length of the body.

● **Vertebrates** In living vertebrates, the notochord develops into a backbone (spine or vertebral column) made up of units called vertebrae. The vertebrate muscular system that moves the body consists primarily of muscles that are arranged in mirror image on both sides of the backbone or notochord.

● **Gnathostomes** These fish have jaws, as opposed to hagfish and lampreys (agnathans or jawless fish), which lack proper jaws. Gnathostomes have gills (breathing apparatus) opening to the outside through slits, and fins that include those arranged in pairs, such as the pectoral (shoulder) fins.

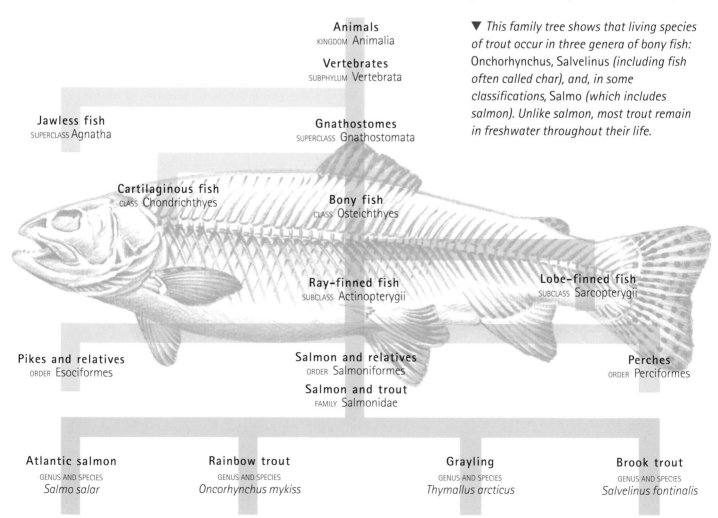

▼ *This family tree shows that living species of trout occur in three genera of bony fish: Onchorhynchus, Salvelinus (including fish often called char), and, in some classifications, Salmo (which includes salmon). Unlike salmon, most trout remain in freshwater throughout their life.*

Animals
KINGDOM Animalia

Vertebrates
SUBPHYLUM Vertebrata

Jawless fish
SUPERCLASS Agnatha

Gnathostomes
SUPERCLASS Gnathostomata

Cartilaginous fish
CLASS Chondrichthyes

Bony fish
CLASS Osteichthyes

Ray-finned fish
SUBCLASS Actinopterygii

Lobe-finned fish
SUBCLASS Sarcopterygii

Pikes and relatives
ORDER Esociformes

Salmon and relatives
ORDER Salmoniformes

Perches
ORDER Perciformes

Salmon and trout
FAMILY Salmonidae

Atlantic salmon
GENUS AND SPECIES
Salmo salar

Rainbow trout
GENUS AND SPECIES
Oncorhynchus mykiss

Grayling
GENUS AND SPECIES
Thymallus arcticus

Brook trout
GENUS AND SPECIES
Salvelinus fontinalis

● **Bony fish** Trout and salmon belong to the class Osteichthyes (bony fish)—the major group that includes more than 95 percent of all fish. Bony fish, as their name implies, have a skeleton of bone. In contrast, members of the class Chondrichthyes (cartilaginous fish, such as sharks, skates, and rays) have a skeleton made of cartilage.

● **Ray-finned fish** Almost all bony fish, including trout and salmon, belong to the subclass Actinopterygii (ray-finned fish). The major feature that distinguishes them from the eight species of the subclass Sarcopterygii (fleshy-finned fish) is the presence of bony rays that support thin, flexible fins.

● **Salmon and their relatives** These fish are all members of the order Salmoniformes, which includes about 140 species in six families. All members are found in temperate freshwater. About 10 species are Northern Hemisphere smelt (family Osmeridae); 12 species are northern icefish (Salangidae); six species are Southern Hemisphere smelt and grayling (Retropinnidae); and 40 species are southern whitebait, galaxiids, and pelladillos (Galaxiidae). A further 70 species of salmonids (family Salmonidae) include char, northern salmon, trout, and whitefish. Finally, the unique salamander fish of Australia belongs in a family of its own (Lepdogalaxiidae). The salamander fish is able to survive in a moist burrow for months at a time when its usual river habitat dries out.

▲ *The brown trout is native to western Asia and Europe but has been introduced to North America, Australia, New Zealand, and several other countries.*

All salmoniforms have two bones in the top jaw, the maxilla and premaxilla, which form the edge of the upper mouth. Most salmoniforms have a small, fleshy adipose fin behind the dorsal fin. Other fins are supported by bony rays and lack spines. Salmoniforms are relatively primitive examples of ray-finned fish.

● **Salmon, trout, char, and northern whitefish** Most members of the family Salmonidae, especially salmon and trout, are slender, with smallish fins and a forked tail. Their scales are small and rounded, and the mouth has single rows of backward-pointing teeth. Salmonids are opportunistic feeders, taking almost any small aquatic animal, including other fish. Their digestive system is simple, reflecting their high-protein diet. Some of the species, especially Pacific salmon and Atlantic salmon, are anadramous: they hatch in freshwater but grow to adulthood in the sea before returning to freshwater to breed. Those species in the genera *Oncorhynchus* and *Salmo* that always migrate from freshwater to the sea and back again are called salmon. Those species in the same genera that generally remain in freshwater throughout their life are called trout. This simple distinction is blurred because some populations of brown trout and rainbow trout do migrate between freshwater and the sea as salmon do.

FEATURED SYSTEMS

EXTERNAL ANATOMY The trout's streamlined body is protected by a covering of small, bony scales. *See pages 1282–1284.*

SKELETAL AND MUSCULAR SYSTEMS The trout swims by passing S-shaped waves back along its trunk and tail. *See pages 1285–1286.*

NERVOUS SYSTEM Vision, smell, hearing, and a vibration-detecting lateral line system are used by trout to home in on prey and avoid predators. *See pages 1287–1288.*

CIRCULATORY AND RESPIRATORY SYSTEMS Trout extract oxygen from the water using gills that employ an efficient countercurrent system. *See page 1289.*

DIGESTIVE AND EXCRETORY SYSTEMS Trout have a simple gut that reflects their meat-rich diet. Chloride cells in the gills pump ions into the fish when in freshwater, or out of the body when the fish is in seawater. *See pages 1290–1291.*

REPRODUCTIVE SYSTEM Eggs are fertilized externally, and the hatchlings rely on their yolk supplies for two to three weeks, until the fish are developed enough to feed for themselves. *See pages 1292–1293.*

External anatomy

CONNECTIONS

COMPARE the cycloid scales of the rainbow trout with the placoid scales of a *HAMMERHEAD SHARK* or *STINGRAY*.

COMPARE the color-producing cells of a rainbow trout with those of a *JACKSON'S CHAMELEON*. Those of a chameleon enable the lizard to change color rapidly, whereas the male rainbow trout changes color only slowly.

Trout, like most bony fish, are streamlined for moderately fast swimming. Water is about 800 times denser than air, so water provides much more support than air. However, water also offers more resistance to an animal moving through it. Thus trout and other types of fish that swim quickly have a streamlined body. This shape reduced drag, which is the resistance to forward movement generated by water turbulence and by friction with the body surface. Trout, like many fish, are fusiform, or torpedo-shaped.

Scales

Scales embedded in the skin provide a flexible protective covering that enables the fish to flex its body for propulsion. The scales are circular, a form called cycloid, and overlap one another like the shingles on a roof. Scales are replaced if lost or damaged. Scars deposited in the scales as they grow enable scientists to assess the age of a fish.

CLOSE-UP

The skin

A trout's skin has two layers. The outer layer, the epidermis, contains mucus-secreting cells and other protective cells. Thick, sticky mucus aids streamlining and protects the skin against invasion by bacteria. The layer beneath the epidermis, the dermis, is richly supplied with blood vessels and nerves. Scales grow from this layer, and parts of the lateral line system are embedded in the dermis. The dermis also contains sensory receptors that make the skin highly sensitive to touch.

The coloration of trout, as in many other fish, is a compromise between the needs for camouflage and communication. Trout need to be well camouflaged to avoid predators and, to

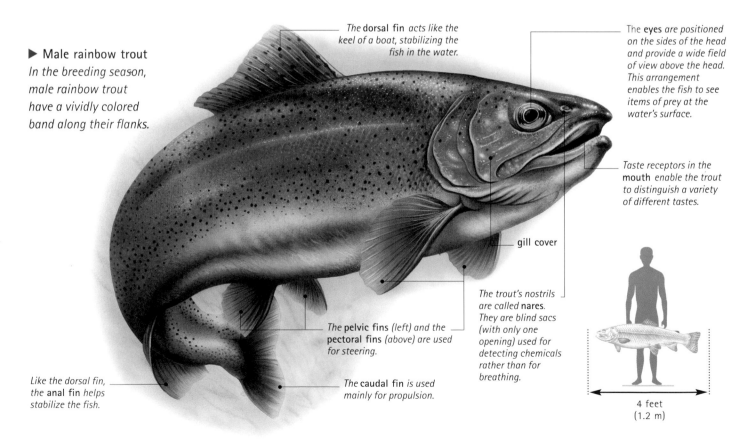

▶ **Male rainbow trout**
In the breeding season, male rainbow trout have a vividly colored band along their flanks.

The **dorsal fin** *acts like the keel of a boat, stabilizing the fish in the water.*

The **eyes** *are positioned on the sides of the head and provide a wide field of view above the head. This arrangement enables the fish to see items of prey at the water's surface.*

Taste receptors in the **mouth** *enable the trout to distinguish a variety of different tastes.*

gill cover

The trout's nostrils are called **nares**. *They are blind sacs (with only one opening) used for detecting chemicals rather than for breathing.*

Like the dorsal fin, the **anal fin** *helps stabilize the fish.*

The **pelvic fins** *(left) and the* **pectoral fins** *(above) are used for steering.*

The **caudal fin** *is used mainly for propulsion.*

4 feet
(1.2 m)

1282

a lesser extent, to avoid being seen by potential prey. Rainbow trout are typically bluish green to brown on the back and sides, and white or yellow on the underside. Numerous black spots extend over the back and sides and onto the dorsal and tail fins. However, trout also need to attract potential mates of their own species. The coloration of a rainbow trout changes with age, with season, and with environmental conditions. For example, males that are ready for spawning are more vividly colored and usually display a pink, red, or lilac band along each flank. Coloration is produced by pigments in specialized cells, called chromocytes, lying

COMPARATIVE ANATOMY

Fish scales

Scales provide a covering on the body surface that protects the underlying tissues against physical damage while maintaining the body's streamlining. Living fish usually have only one of five possible types of scales. Cartilaginous fish, such as sharks and rays, have scales made of dentine and a substance similar to tooth enamel. Their scales are called placoid: they are sharp and pointed, with the widest part embedded in the skin. Primitive bony fish also have scales made of dentine and enamel-like substances, but the scales are flattened rather than spiny. In gars, the scales are ganoid: they are diamond-shape, interlocking, and anchored to one another by protein fibers. In coelacanths, the scales are called cosmoid and are made of dentine overlaid with a type of enamel called cosmine. In more modern bony fish, the scales are thin, flexible, overlapping, and made of bone. One end is embedded in the skin, and the other end is exposed. Fish such as perch and carp have ctenoid scales, in which the exposed surface is rough or spiny. Trout and salmon have cycloid scales in which the exposed part is smooth.

embedded part of scale

embedded part of scale

embedded part of scale

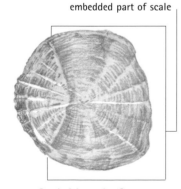

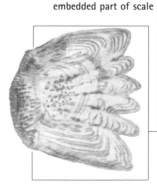

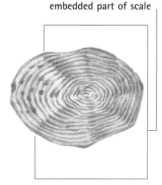

Cycloid scale from
common carp

Ctenoid scale
from perch

Cycloid scale from
rainbow trout

Two species become one

Until the 1980s, scientists thought that the rainbow trout of North America (*Salmo gairdneri*) was a different species from the native trout of Siberia in Russia (*Oncorhynchus mykiss*). Using fossil evidence, scientists came to realize that the two species were the same and that both were more similar to Pacific salmon and trout (genus *Oncorhynchus*) than to Atlantic species (genus *Salmo*). The name of the North American rainbow trout was changed to that of the Russian species in 1989.

toward the base of the dermis. Different kinds of chromocytes contain red, yellow, brown, or black pigments, and the combinations of these overlaying one another produce other tones. Some cells called iridiocytes contain crystals of a substance called guanin, which acts like a reflective mirror. Iridiocytes lying beneath chromocytes produce a characteristic metallic sheen on the flanks of the trout.

Two types of fins

Like most other fish, trout have two types of fins. Those that lie in the midline, called median fins, are the tail, or caudal, fin; the dorsal, or upper fin on the back; and the anal fin, which lies on the underside, or ventral surface. In addition, trout and salmon have a small, fleshy adipose fin behind the dorsal fin. The caudal fin is used mainly for propulsion in swimming but can also act as a rudder for steering. The dorsal and anal fins act like the

▼ *The name* brown trout *does not adequately describe the colors found in this species. The gray-blue body is covered with spots that range from black on the upper body to orange on the sides. The body is pale brown below. Overall colors may vary, however, according to location.*

keel of a boat, stabilizing the fish and helping prevent it rolling as it swims. Trout, like most fish, also have two sets of paired fins, with one fin of each pair on either side of the midline. The two pectoral (shoulder) fins are located in front of the two pelvic fins. The paired fins provide fine control for steering and can be angled to make the fish ascend or dive as it swims. All fins, except the adipose fin, are supported by rays of bone. The caudal fin is large and slightly forked and enables the trout to accelerate rapidly to capture prey and evade predators.

Blind-ended nares

The nostrils are technically called nares, and, unlike the nostrils of many land-living vertebrates they have only one opening and are separate from the respiratory system. The lining of the olfactory (smell-sensing) sac is highly folded to accommodate more sensory cells. These cells are highly sensitive to odors in the water and can detect substances at concentrations of only 1 part in 80 billion, which is equivalent to less than a teaspoonful dissolved in the water of an Olympic-size swimming pool.

Large eyes

The trout's eyes are moderately large and are positioned on the upper portion of the sides of the head, providing a wide field of view above the fish. A large proportion of the trout's diet is insects that drop onto the water surface, insect larvae, and fish traveling in midwater.

Skeletal and muscular systems

Like other bony fish, trout and salmon have a skeleton that is made up of a skull, jaws, and backbone, as well as bony rays that support the fins. In the rainbow trout the backbone contains numerous vertebrae that move slightly relative to one another, allowing the body to flex. The first half of the vertebral column has points of attachment for ribs, which extend around the body cavity and provide some support and physical protection for the soft internal organs contained within. In the head region, the skull provides housing and protection for the brain, eyes, ears, and olfactory organs. The skeleton surrounding the mouth and gills consists of upper and lower jaws and support for four gills on either side. A bony cover, the operculum, protects each gill cavity.

In trout, as in most bony fish, the pectoral fins are attached to the skull by a pectoral girdle, whereas the dorsal, pelvic, and anal fins are anchored into muscle blocks. Complex arrangements of muscles control movements of the jaws and gill apparatus. Six separate muscles move each eye. The fins each have their own muscular system.

Swimming through water demands large body muscles, and typically 50 to 60 percent of the mass of a salmon or trout is made up of swimming muscles. These are arranged in W-shaped muscle blocks, or myotomes, on either side of the backbone, with the points of the W pointing toward the tail. Each muscle block is divided into two parts by connective tissue: the upper (epaxial) part above the spinal cord and the lower (hypaxial) part below are separated by a boundary called the horizontal septum.

Most fish swim by throwing the body and tail into an S-shaped wave that passes backward along the body, becoming more enhanced the farther back it travels. The bending is produced by the myotomes on both sides of the vertebral column. As the myotome on one side of the body contracts, that on the other relaxes, so the body bends in the direction of muscle contraction. When muscle blocks contract, one after the other in a synchronized fashion, a propulsive force is produced that pushes the fish forward. The muscle blocks are composed of striated skeletal muscle that is controlled by nerves connected to the central nervous system (CNS).

COMPARE the buoyancy of a trout's swim bladder with that of a *STINGRAY*'s liver. Cartilaginous fish, such as stingrays, do not have a swim bladder. Instead, they have a large, oil-rich liver that provides buoyancy.

▼ Rainbow trout
The skeletal system of a rainbow trout provides support for the body and fins, as well as protection for soft organs such as the brain and intestines.

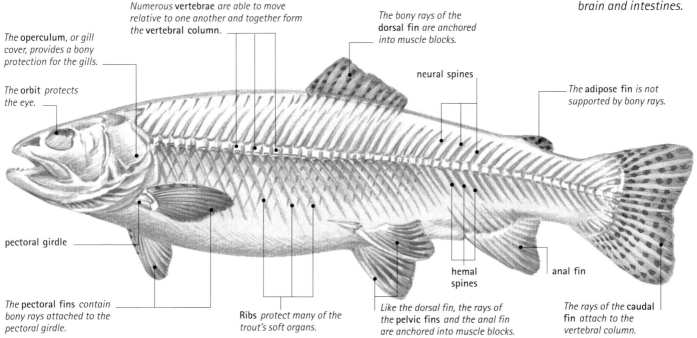

Numerous **vertebrae** *are able to move relative to one another and together form* the **vertebral column.**

The bony rays of the dorsal fin *are anchored into muscle blocks.*

The **operculum,** *or gill cover, provides a bony protection for the gills.*

The **orbit** *protects the eye.*

neural spines

The **adipose fin** *is not supported by bony rays.*

pectoral girdle

The **pectoral fins** *contain bony rays attached to the pectoral girdle.*

Ribs *protect many of the trout's soft organs.*

hemal spines

Like the dorsal fin, the rays of the **pelvic fins** *and the anal fin are anchored into muscle blocks.*

anal fin

The rays of the **caudal fin** *attach to the vertebral column.*

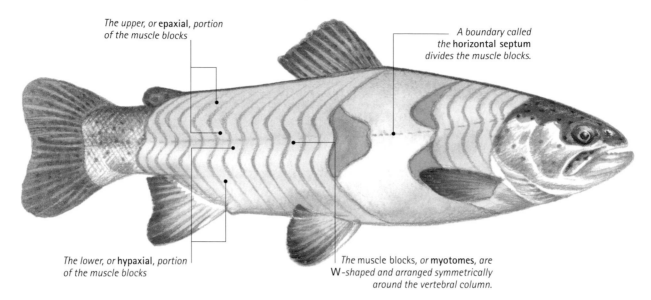

The upper, or **epaxial**, portion of the muscle blocks

A boundary called the **horizontal septum** divides the muscle blocks.

The lower, or **hypaxial**, portion of the muscle blocks

The muscle blocks, or **myotomes**, are W-shaped and arranged symmetrically around the vertebral column.

▲ **Rainbow trout**
Blocks of muscle, called myotomes, contract along the body to propel the fish through water.

An air-filled sac called the swim bladder provides bony fish with buoyancy, enabling them to maintain their vertical position in the water with minimal energy consumption. The swim bladder of the trout, unlike that of most other bony fish, is connected to the gut by a tube called the pneumatic duct. This is a primitive feature. To increase the volume of air in the swim bladder, the fish must rise to the surface and gulp air, which it passes from the gut to the swim bladder. This mechanism limits trout and salmon to living in shallow water.

IN FOCUS

Red and white muscle

If a trout is cooked, its flank muscle shows up as partly dark red or brown but mostly pink or white. The red muscle is concentrated along each flank around the boundary between the epaxial and hypaxial muscles. These muscles are unusually rich in blood vessels but are also rich in myoglobin, an oxygen-carrying pigment similar to the hemoglobin found in red blood cells. Myoglobin enables red muscle to work for long periods at high capacity without tiring, as when the fish maintains its position in a fast-flowing current of water.

The flank's white muscle has a smaller blood supply and lower levels of myoglobin compared with red muscle. White muscle is able to contract more powerfully than red muscle. White muscle gives shorts of bursts of high speed but tires relatively quickly. It comes into use to produce a fast burst of speed for capturing prey or evading a predator.

▶ **MUSCLE SECTION OF TROUT**
Blocks of muscle surround the vertebral column and body cavity. The muscle is a mixture of non-tiring red muscle and white muscle, which is used for bursts of speed.

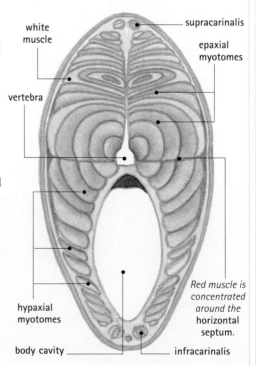

white muscle

supracarinalis

epaxial myotomes

vertebra

Red muscle is concentrated around the horizontal septum.

hypaxial myotomes

body cavity

infracarinalis

Nervous system

The nervous system of a trout has the same arrangement as that of other bony fish. The central nervous system (CNS) consists of a brain and spinal cord. The CNS is connected via nerves of the peripheral nervous system (PNS) to sensory organs and to responsive structures (effectors) such as muscles.

The brain is broadly divided into three regions, as in other vertebrates: the forebrain, midbrain, and hindbrain. In amphibians, reptiles, and birds and mammals especially, the forebrain has the central coordinating role for major sensory organs and voluntary activities. In fish, however, this coordinating role is spread between different parts of the brain. The forebrain coordinates responses to scent, and the first pair of cranial nerves, the olfactory nerves, arise from it. The forebrain also controls territorial, courtship, and spawning behaviors.

The midbrain is responsible for controlling responses to visual stimuli, and for this purpose it contains two bulbous regions, the optic lobes. Just behind the optic lobes lies the cerebellum, which is part of the hindbrain. This coordinates body orientation and major body movements. The medulla oblongata, lying behind the cerebellum, controls automatic (involuntary) actions, such as the heart rate.

A range of senses

Trout locate their prey by a combination of sight, smell, hearing, and other forms of vibration detecting. The eyes of trout and bony fish typically lack eyelids. The eyes focus by

movement of the lens back and forth within the eye chamber, rather than by changing the shape of the lens as in mammals. Trout, like other shallow-water fish, can see distances of some 16 to 33 feet (5–10 m) and can perceive objects both above and below the water surface. The light-detecting layer at the back of the eye, the retina, contains highly light-sensitive rod cells, which detect levels of light as shades of gray; and less light-sensitive cone cells, which perceive color.

A pair of nares located at the tip of the fish's snout sample the watery environment for dissolved chemicals, producing the fish's sense of

▲ *When a trout leaps out of the water its movement is coordinated by the nervous system, which controls the contractions of the muscles and monitors the position of the trout relative to its surroundings.*

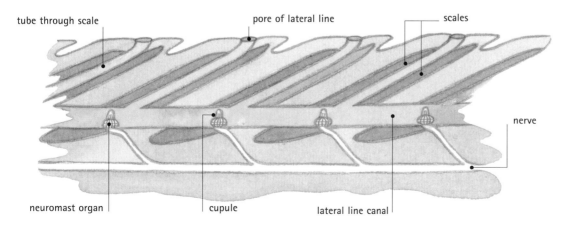

tube through scale　　pore of lateral line　　scales

nerve

neuromast organ　　cupule　　lateral line canal

◄ LATERAL LINE SYSTEM
The lateral line system detects vibrations and pressure changes in the water. These disturbances travel through pores in the fish's scales and along gel-filled canals to sensory organs called neuromasts.

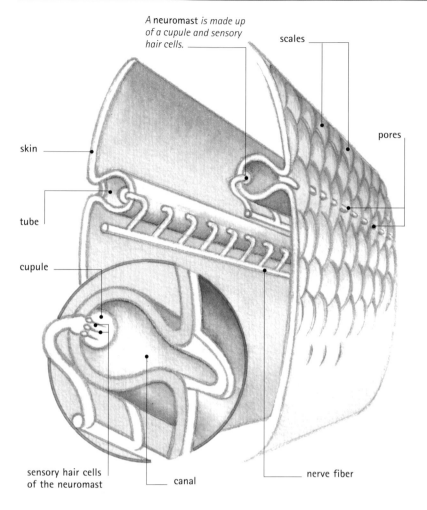

A **neuromast** *is made up of a cupule and sensory hair cells.*

skin

tube

cupule

sensory hair cells of the neuromast

canal

scales

pores

nerve fiber

▲ **NEUROMAST**

Neuromasts are part of the lateral line system, which runs around the head and along the sides of the trout. Sensory organs called neuromasts detect movements in water, allowing the fish to sense not only its prey but also its predators.

smell. The nares lead to blind-ended olfactory sacs that are lined with chemosensory cells. The sense of smell is acute, and trout can detect dissolved chemicals at levels well below one part per billion.

Taste receptors lie in the outer cell layer, the epidermis; inside the mouth; and on the gill rakers. These receptors allow the trout to distinguish a variety of tastes—including bitter, sweet, acid, sour, and salty—helping it to confirm whether or not a prey item is safe to eat. The trout can also taste the water it breathes in.

The trout's two ears are embedded in bony capsules at the rear of the skull. They have no direct connection to the surroundings, unlike the ears in most land-living vertebrates. Nevertheless, fish ears detect vibrations that travel through the water, and the brain processes this information, interpreting the vibrations as sound. The hearing part of each ear is in two pouches, called the saccule and lagena, of the inner ear. They correspond to the cochlea of land-living vertebrates. The

trout's swim bladder acts like a drum to amplify the vibrations passing through the water, greatly increasing the fish's hearing sensitivity.

Sense of balance

The ears are also responsible for the fish's sense of balance. This is achieved by a system of three fluid-filled canals oriented roughly at right angles to one another, called the semicircular canals; and three chambers called the utriculus, sacculus, and lagena. Each canal has a bulbous region, called the ampulla, which contains a gel-like cupule connected to the hairs of sensitive cells. The three chambers also have similar arrangements but with chalky granules called otoliths instead of cupules. When fluid moves through the system, as a result of the body's accelerating, decelerating, or changing direction, the cupules or otoliths are displaced and pull on the hairs attached to nearby sensory cells, which send nerve impulses to the brain, where the stimuli are interpreted. The combination of stimulations to sensory cells enables the fish to monitor its orientation and pattern of body movement.

Lateral line system

Like most fish, trout detect vibrations in the water using a lateral line system. This is an extensive system of small canals running along the middle of each flank and opening at the body surface through tiny pores that pass through scales. The system also branches into a series of canals and pores in the head.

The lateral line system detects vibrations and pressure variations in the water when they cause the gel-like fluid in the pores and canals to move. This system triggers sensory organs within the canal system, called neuromasts. As noted above, each neuromast consists of a gel-like structure, called the cupule, connected to a group of sensory cells by sensory hairs. Vibrations displace the cupule, which pulls on the sensory hairs, triggering the sensory cells to send electrical impulses to the brain, where the stimulus is interpreted. This mechanism is very similar to that found in the ampullae of the inner ear. Using the lateral line system, the trout can sense disturbances in its surroundings, such as those produced by potential prey or predators.

Circulatory and respiratory systems

Like other bony fish, such as sailfish, trout have a single-circuit blood circulation system. Blood is pumped at high pressure from the heart through the arteries. In the gills, the arteries divide into much smaller blood vessels, called capillaries, which pick up oxygen and release carbon dioxide. The blood vessels that exit from the gills merge into larger vessels, before dividing again into capillaries to supply other tissues around the body. The blood then returns to the heart at lower pressure through the system of veins.

Like most other fish, trout use gas exchange across the gills to gain the oxygen they need for respiration and to get rid of carbon dioxide. To create the water flow for gas exchange, trout take in water through the mouth and force it out through gill slits. They achieve this force by enlarging first the mouth cavity and then the gill cavities, and then constricting both in sequence, so water flows from the closed mouth toward the gill slits. This breathing method is highly coordinated and efficient, producing an almost continuous flow of water across the gills. Four gills on each side form a curtain that separates the mouth cavity from the gill cavities. To pass from one cavity to the other, water must pass through the gills.

CLOSE-UP

Gill structure and function

Each gill of a bony fish is supported by a bony arch called a gill arch from which two rows of bony extensions called gill rakers extend forward. The gill rakers prevent large objects from passing through and damaging the gills, and they also strain off plankton and other small food items, which can be swallowed. The gas exchange surface of each gill is made up of numerous gill filaments, rather like plates stacked one on top of the other. The surface of each gill filament is itself folded into platelike structures called lamellae. The combination of filaments and lamellae creates a very large surface area across which gases are exchanged between the fish's blood and the flow of water. Each lamella is richly supplied with blood vessels. These pick up oxygen from the moving water while releasing waste carbon dioxide into it. At the gill lamellae, water and blood flow in opposite directions so that newly arriving water meets blood that is about to exit from the gills. This mechanism, called countercurrent flow, maximizes the rate of gas exchange by ensuring steep diffusion gradients for the dissolved gases. More than 50 percent of the oxygen in the water flow is taken into the blood.

▶ GAS EXCHANGE ACROSS THE GILLS

To pick up fresh oxygen and get rid of waste carbon dioxide, trout force water across their gills. The gills have a large surface area and are richly supplied with blood vessels. Because water and blood flow in opposite directions (countercurrent flow), the amount of oxygen extracted from the water is maximized.

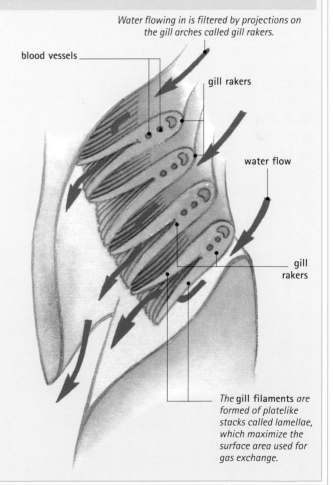

Water flowing in is filtered by projections on the gill arches called gill rakers.

blood vessels

gill rakers

water flow

gill rakers

The gill filaments are formed of platelike stacks called lamellae, which maximize the surface area used for gas exchange.

Digestive and excretory systems

COMPARE the length of a trout's digestive tract with that of the **RED DEER**. The trout has a short digestive tract typical of animals with a meat-rich diet, whereas the red deer has a very long, complex digestive tract typical of a plant-eating animal.

▼ Rainbow trout
The digestive process begins in the mouth, where the teeth crush ingested prey. Food is further broken down in the acidic environment of the stomach before passing into the short intestine, from which most nutrients are absorbed into the blood.

Most predatory fish use one of two feeding strategies. Some are suction feeders, with a small mouth that sucks in water like a pipette when the mouth cavity is expanded. The suction effect enables the fish to draw in water and with it small organisms from a distance of up to a few inches. The alternative feeding method is to have a large mouth and engulf items of prey by swimming rapidly to overtake and swallow them. This method is called ram feeding. Trout and salmon have a mouth of moderately large size and tend to use the ram feeding method. However, they can also use the suction effect to draw in small invertebrates from the bottom of a river or lake.

Digestive tract

Trout and salmon have teeth, which the fish use to crush their prey before swallowing it. The lining of the mouth cavity and the esophagus, or gullet, that follows it contain mucus-secreting cells that coat the food with mucus to allow it to pass through the gut.

In common with most other fish that eat mainly flesh, the trout has a fairly short gut. The esophagus empties into the stomach. There, physical and chemical digestion—the

Chloride cells

Remarkable cells in the gill epithelium enable trout to pump salts into or out of the body to maintain the balance of salt and water. These cells are densely packed with mitochondria—cell organelles that are the site of energy production. The presence of so many mitochondria confirms that this pumping action requires large amounts of energy. In freshwater, these chloride cells pump ions such as chloride (Cl^-) and calcium (Ca^{2+}) into the body. In seawater, these same cells pump chloride (Cl^-) and sodium (Na^+) ions out of the body. When a trout or salmon is traveling from freshwater to the sea, it rests for a few days in brackish water (a mixture of seawater and freshwater) while its chloride cells multiply and their pumping action gradually changes direction.

breakdown of food—begins. The inside of the stomach is extremely acidic, with a pH of 2 to 3, which kills most swallowed bacteria. The

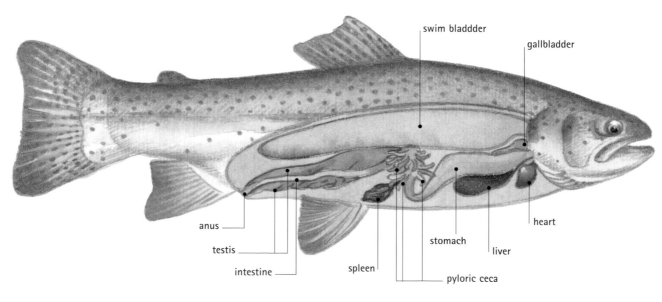

swim bladdder

gallbladder

anus

testis

intestine

spleen

stomach

liver

heart

pyloric ceca

low pH also provides a suitable environment for the protein-digesting enzyme pepsin, which is secreted by the walls of the stomach. From the stomach, the partially digested food is squirted into the first part of the intestine. The intestine receives a range of digestive enzymes in juice from the pancreas. The pancreatic enzymes gradually break down the various food components—carbohydrates (complex sugars), proteins, and fats—into smaller chemical substances that are absorbed across the lining of the intestine and into the blood. Salmonid fish have blind-ended processes, called pyloric ceca, that emerge from the intestine and increase the area available for digestion and absorption. From the intestines, the absorbed products of digestion travel in the blood to body tissues. There, they are used to make body parts or are broken down to release energy in the process of cellular respiration. Undigested material remaining in the gut is expelled through the anus.

Liver and kidneys

The liver of a trout is relatively large and produces bile that passes into the first part of the intestines, where it helps break down fats. The liver also stores blood sugar in the form of glycogen, which serves as an energy store; and it breaks down toxic substances circulating in the blood. The pancreas releases the hormone insulin, which helps mobilize the liver to store glucose as glycogen. The liver acts as a "clearinghouse" for products of digestion and stores fat and vitamins A and D, as well as glycogen.

The trout's two kidneys are large, dark red organs in the upper part of the abdominal cavity. As in many other bony fish, the kidneys filter the blood and remove dissolved waste substances that are expelled from the body in urine. Like most freshwater fish (but unlike most marine fish), trout do not have a urinary bladder for storing urine. Instead, the urine is expelled as fast as it forms.

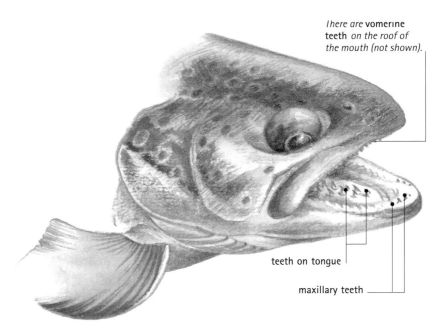

There are **vomerine teeth** on the roof of the mouth (not shown).

teeth on tongue

maxillary teeth

▲ TEETH
Rainbow trout
Trout have different types of teeth: teeth on the tongue, maxillary teeth growing from the jaw, and backward-pointing vomerine teeth on the roof of the mouth.

Migration between freshwater and sea water

Lake-dwelling trout migrate into tributary streams to spawn. Some populations of river trout spend their adult life at sea, returning to the places where they hatched to spawn as adults. To leave freshwater and enter sea water, fish experience a dramatic change in salinity. They leave salt-poor freshwater to enter salt-rich sea water. The osmotic conditions of freshwater are entirely different from those of sea water. In freshwater, the concentration of salts within the fish's body is much higher than that in the watery environment, and water enters the fish by osmosis. Trout, like most other freshwater fish, expel large quantities of dilute urine to get rid of excess water. They also pump ions into their body across the gills, using special chloride cells, to maintain the high salt concentration within the body. In sea water, osmotic conditions are reversed. The concentration of salts within the fish's body is less than that in the surroundings. Water leaves the fish by osmosis, and the fish conserves water by producing highly concentrated urine. Trout and salmon drink sea water and pump the excess ions out across their gills using chloride cells, which pump in the opposite direction to that in freshwater.

Reproductive system

CONNECTIONS

COMPARE fertilization in the trout with that in the *STINGRAY*. Trout practice external fertilization, in which sperm from the male mix with the female's eggs in the water. Stringrays, however, practice internal fertilization, in which the male inserts his sperm directly into the female.

Male trout have a pair of sperm-producing organs, the testes, in the lower abdomen. Females have two egg-producing organs, the ovaries. At spawning time, females release hundreds of small eggs. These are fertilized in the water by microscopic, tadpolelike sperm, released in their millions by males.

Rainbow trout are found in many different kinds of habitats, from huge lakes to tiny mountain streams. Some live their entire life in landlocked environments, and others migrate to and from the sea. It is thus impossible to generalize about all the life cycle habits of this species. In small streams, a 12-inch (30-cm) fish is an elderly giant, whereas in a large lake or river, a fish of that size is barely three years old and could grow to more than 3 feet (90 cm) long in old age.

The rainbow trout of temperate streams and rivers usually spawn in spring or later, depending on location and water temperature. Males, called cock fish, usually arrive on the spawning grounds before females. The males

▶ **Rainbow trout**
Female trout have two ovaries, in which hundreds of eggs are produced. Males have two testes, which produce sperm.

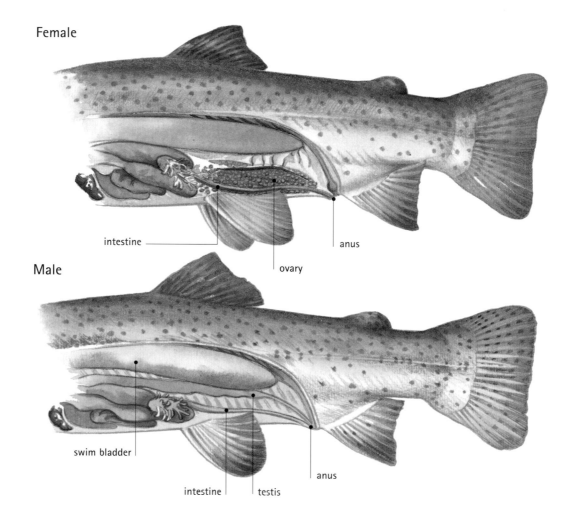

Female

intestine

ovary

anus

Male

swim bladder

intestine

testis

anus

Death after spawning

Pacific salmon and trout that migrate to and from the sea die after spawning and therefore spawn only once in their lifetime. Their decaying bodies add large quantities of nutrients back into the rivers in which the fish began their life. Rainbow trout and other trout that remain in freshwater all their life usually spawn several times.

Hatching and development

In water temperatures between 50°F and 59°F (10–15°C), the eggs hatch after 21 to 28 days. The young fry remain for their first 16 to 20 days in the gravel beds of their spawning grounds, obtaining their all the nutrients they need from the yolk in their yolk sacs. The fry then leave the spawning grounds and form free-swimming shoals. They feed first on the aquatic larvae of midges and mayflies brought to them by the current.

Within a few weeks, however, the fry scatter and take up feeding stations, with each fish separated from the next by several feet. At this small size, the fish are called parr. When slightly older, they usually migrate away from the spawning grounds to wider stretches of river or lake, or even the sea, where they will feed and grow into adults before returning to spawn two to three years later.

As adults, trout continue to feed on aquatic insects, crustaceans, and small fish, in addition to small animals such as insects that drop onto the water's surface. Few trout survive beyond seven years old.

TREVOR DAY

FURTHER READING AND RESEARCH

Mayden, R. L. 1992. *Systematics, Historical Ecology, and North American Freshwater Fishes.* Stanford University Press: Stanford, CA.

Moyle, P. B., and J. J. Cech. 2000. *Fishes: An Introduction to Ichthyology* (4th ed.). Prentice Hall: Upper Saddle River, NJ.

Paxton, J. R., and W. N. Eschmeyer (eds). 1998. *Encyclopedia of Fishes* (2nd ed.). Academic: San Diego, CA.

develop a hooked jaw and brighter colors at spawning time, and the fish compete with one another for the best stretches of river for spawning. Larger fish drive away smaller fish from their territories. When females arrive, they excavate characteristic hollows, called redds, on the riverbed. They do this by turning on their side and beating their tail to produce a powerful wash that sweeps away gravel. The female typically releases several hundred eggs, which are fertilized by the locally dominant male, who squirts milt (a fluid rich in sperm) over the eggs as they are released. The female loosely covers the fertilized eggs with gravel to protect them and help keep them from being washed away by the current. The process is repeated several times over a few days, sometimes with different males. Finally the exhausted female leaves while the male often guards the eggs and later the hatchlings for their first few days.

▲ *Rainbow trout hatch 21 to 28 days after fertilization. The large yellow yolk sac provides the fry with the nutrients they need for the first days of their life.*

Index

Page numbers in **bold** refer to main articles; those in *italics* refer to picture captions, or to names that occur in family trees.